The Central Narmadā Valley

The Central Narmadā Valley

A Study in Quaternary Palaeontology and Allied Aspects

G.L. Badam
Senior Consultant
Indira Gandhi National Centre for the Arts
New Delhi

Appendices by
Rajeev Patnaik, Arati Deshpande Mukherjee,
Anupama Kshirsagar, G.L. Badam and Vijay Sathe

Indira Gandhi
Rashtriya Manav Sangrahalaya
Bhopal

D.K. Printworld (P) Ltd.
New Delhi

Cataloging in Publication Data — DK
[Courtesy: D.K. Agencies (P) Ltd. <docinfo@dkagencies.com>]

Badam, Gyani Lal, 1940–
The Central Narmadā Valley : a study in quaternary palaeontology and allied aspects / by G.L. Badam.
xxviii, 210 p. 29 cm.
Includes bibliographical references (p.)
Includes index.
ISBN 8124604258 (10 digit)
ISBN 9788124604250 (13 digit)

1. Vertebrates, Fossil — India — Narmada River Valley. 2. Paleontology — India — Narmada River Valley. 3. Paleontology — Quaternary. 4. Geology, Stratigraphic — Quaternary. 5. Excavations (Archaeology) — India — Narmada River Valley. 6. Narmada River Valley (India) — Antiquities. I. Title.

DDC 560.456 409 543 22

ISBN 81-246-0425-8 (10 digit) (Hardbound)
ISBN 978-81-246-0425-0 (13 digit) (Hardbound)
First published in India in 2007

Published by:
Indira Gandhi Rashtriya Manav Sangrahalaya
Post Bag no. 2, Shamla Hills,
Bhopal - 462 013 (Madhya Pradesh)
Phones: (0755) 266 1319, 423 2058
Fax: 91-755-266 1458
E-mail: homosapien@sancharnet.in
Website: www.museumofmankindindia.gov.in

and

D.K. Printworld (P) Ltd.
Regd. Office: 'Srikunj', F-52, Bali Nagar
Ramesh Nagar Metro Station
New Delhi - 110 015
Phones: (011) 2545 3975; 2546 6019; *Fax*: (011) 2546 5926
E-mail: dkprintworld@vsnl.net
Website: www.dkprintworld.com

Printed by: D.K. Printworld (P) Ltd., New Delhi

to my dearest ones

Lalita

Priya, Anand & Aryan

Puneet & Sheeba

Foreword

In recent times, a great deal of literature has been published about the Narmadā Valley. For several reasons, academic and religious, the Narmadā Valley has been the subject of numerous expeditions, seminars, debates, publications (both scientific and popular) and discussions, sometimes even in the Parliament.

The Narmadā originates from the western flank of the Maikala Range near Amarkaṅṭak in Madhya Pradesh, and flows through a rift valley (geologically known as Narmadā lineament) across the middle of the Indian subcontinent. Almost 80 per cent of the total length of Narmadā (1300 km) lies in Madhya Pradesh from where the Narmadā flows through Gujarat on its way to the Arabian sea. Narmadā has preserved some of the finest treasures of archaeological, geological, historical and religious importance. Like the Ganges, Narmadā is believed to have descended to earth from Śiva's *jaṭā* (top knot). Hindus believe that while it requires a bath in the Ganges to wash away one's sins, the mere sight of Narmadā is sufficient to achieve this goal. That may be the reason why thousands of devotees go on what is known as *Narmadā Parikramā*, from the source at Amarkaṅṭak to the Gulf of Cambay and back.

It is generally believed that Narmadā is not just a river, but a culture, as many mythical legends, folklore and folk songs are associated with it. Several tribes live along the banks of the Narmadā and have been a subject of detailed study by anthropologists, sociologists and non-Governmental Organizations (NGOs). Geological and archaeological data have been collected by foreign and Indian explorers and scholars during the last 150 years or so. Interpretation of these data has thrown a flood of light on the river behaviour, mode of deposition of sediments, chronology, palaeo-environment, man-land relationship and other aspects of the past.

During the last two decades, Narmadā came to be included in the Fossil Hominid map of the world because of the discovery of a partial cranium reportedly of *Homo erectus narmadiensis* by officers of the Geological Survey of India. Thousands of cultural and faunal material collected from the valley give clues to the presence of the enormous animal wealth and the traces of Early Man during the time span of 0.7 million years to 25,000 years ago. In recent years, Narmadā has attracted worldwide attention due to construction of major Sardar Sarovar, Indira Sagar and other dams on the river system. Out of the 30 major, 300 medium and a few thousand minor irrigation and hydroelectric dam projects proposed, some have already been completed. The massive Sardar Sarovar dam, which could make a lot of water available to drought-prone areas of Gujarat and Madhya Pradesh, became a centre of controversy. Groups of people and NGOs opposed the proposal because it would submerge almost 300 villages, render several thousands homeless and destroy valuable forests. Much has been written on this problem especially on the rehabilitation of oustees.

The Indira Gandhi Rashtriya Manav Sangrahalaya (IGRMS) (National Museum of Mankind) is dedicated to the depiction of the story of mankind evolving in time and space. Because of the close proximity of the Museum site to various water bodies like Narmadā and Betwā rivers, one of the major projects undertaken by the Museum is on the history of waters and traditional water harvesting structures, in which the main emphasis has been laid on the revival of age-old water management systems and people's participation for their survival.

The IGRMS has undertaken this effort as part of its larger endeavour to depict the history of human interventions in the earth planet. This endeavour, in turn, is integral to the New Museum Movement, initiated by the IGRMS in the field, for treating floral, faunal and human communities as part of a living museum of biocultural landscape. The new Museum should act as a catalytic agent for the preservation and revitalization of the variety and complexity of this living museum against the engulfing tide of homogenization and technification.

The Narmadā Valley, thus, provides a backdrop for the history of waters and associated biological and cultural history, for making an open-air and indoor exhibition on waters and of select musuem initiatives in the field for the preservation of life enhancing elements in the organic and inorganic communities that people the banks of Narmadā. The book "The Central Narmadā Valley: A Study in Quaternary Palaeontology and Allied Aspects," is the first in a series of volumes in connection with history and management of waters, planned by the IGRMS and is a result of extensive fieldwork and expeditions conducted along the Narmadā Valley. In order to salvage the rich cultural heritage of the Narmadā river valley, the IGRMS approved a project "Taphonomy and Palaeoecology of the Fossil

Vertebrates from the Central Narmadā Valley, Dist. Narsinghpur, Madhya Pradesh, India," which has resulted in the preparation of the present book.

The project was implemented by Dr G.L. Badam, a senior scholar and biostratigrapher (then based at Deccan College Postgraduate and Research Institute, Pune), with a focus on the fossiliferous areas, lying between Jabalpur and Narsinghpur. In collaboration with the IGRMS, Dr Badam conducted surveys and excavations for three field seasons (1992-94) and collected valuable data, bearing upon various aspects of geology, palaeoecology and man-land relationship in the area. Apart from various antiquities, which include stone age tools, fossils, molluscan shells, Dr Badam's team also collected microvertebrate fossils, for the first time, in the entire valley. These are ecologically sensitive and chronologically reliable markers. Thus, the present work is multidisciplinary in nature, incorporating various branches of scientific archaeology. The study has extended the dimensions of understanding of the life of river Narmadā, not only by discovering and reinterpreting various geological, biological and cultural events, but also by analysing the formation processes of the river in its entirety.

The present book is expected to throw considerable light on various aspects discussed above and will prove useful to the students and scholars of river valley cultures in general and to those of the Narmadā Valley in particular. I hope that the sanctity of the river will be respected and the tangible and intangible heritage, geological, palaeontological and archaeological sites will be preserved along the banks of the Narmadā for the benefit of humanity at large.

K.K. Chakravarty

From the Desk of the Director

It gives us immense pleasure to bring out the monograph *The Central Narmadā Valley : A Study in Quaternary Palaeontology and Allied Aspects* written by Dr G.L. Badam, a well known palaeontologist and biostratigrapher. This work is the outcome of a project funded by IGRMS during 1990s when Dr Badam was based at Deccan College, Pune. Since our museum is involved in depicting and exhibiting a story of human kind in time and space, interdisciplinary fieldwork and research have always been encouraged. Such an approach was also followed for the Narmadā Valley of which Dr Badam's work constitutes an important component. It was also felt that the output of such interdisciplinary work on the Narmadā would be helpful in developing the Open Air Exhibition on "River Valley Cultures of India". We congratulate Dr Badam for his in-depth study with a bearing on palaeoecology and man-land relationship in the Central Narmadā Valley. M/s D.K. Printworld (P) Ltd., New Delhi deserve our special thanks for showing their keen interest in publishing this volume. This book is yet another pointer to the success story of our publication policy of public-private partnership. It is sincerely hoped that the scholarly world in general and prehistorians, palaeontologists and ecologists in particular would find the work academically useful.

Bhopal
27th June, 2007

Kishor K. Basa
Director
Indira Gandhi Rashtriya Manav Sangrahalaya

Preface

THE Narmadā, a west flowing river through the middle of the Indian subcontinent, is considered to be a sacred river in India. It has not only preserved rich archaeological, geological and palaeontological material but also important sites of historical and mediaeval ages. The valley has assumed much significance with the discovery of a partial skull of a hominid, *Homo erectus narmadiensis* at Hathnorā (Hoshangabad Dist.) in 1984 by the officers of the Geological Survey of India. Though work on this valley has been carried out for more than 150 years by several foreign and Indian scholars including the present writer, the hominid discovery has infused renewed interest in the study of various aspects of the valley. The recent discovery of microvertebrates is also expected to induce several microfaunal experts to bring this less known fauna to light. For the writer, the valley has been and will always remain a great source of academic satisfaction, because of the rich and varied palaeontological data that it has yielded from time to time.

As is well known, the work for the construction of the Narmada Sagar Dam on the Narmadā river has met with several objections from a section of environmentalists and social activists. However, the dam, when completed will submerge some areas containing archaeological treasures. Though it will not directly affect the main fossiliferous areas under study, the zone of submergence falls partly in the neighbouring districts of Hoshangabad, Harda, Harsud, E. Nimar, W. Nimar, Dhar, Barwani, etc. This necessitated a proper documentation and investigation of the archaeological and palaeontological record in its original geological/sedimentary context, lest the treasures should be lost to posterity for ever.

Thus, the Archaeological Survey of India and the Madhya Pradesh State Dept. of Archaeology and Museums surveyed numerous villages (about 300) and collected a lot of evidence of prehistoric, historic and mediaeval sites, temple sites, sculptures, iron smelting sites, etc. They also formulated plans for salvaging more areas before the region was fully submerged.

The Indira Gandhi Rashtriya Manav Sangrahalaya (IGRMS), Bhopal, took keen interest in the task of collecting and documenting the precious antiquities from the entire valley including parts of the areas in the zone of submergence. To achieve this end and to understand the Pleistocene biological environment and man-land relationship in the area, a project proposal, submitted by the writer, was sanctioned by the IGRMS. In this manner, a collaborative research programme was launched in 1992 between the Deccan College, Pune and the IGRMS, Bhopal. Another project was sanctioned by the IGRMS in favour of a colleague, who worked on an interesting aspect of archaeology (site formation process) of Mehtakheri in Dist. Nimar. It is hoped that, in future, more collaborative projects will be undertaken jointly by the two prestigious institutions, which would further help unravel the mysteries of the prehistoric Madhya Pradesh.

The present book is the outcome of the project entitled "Taphonomy and Palaeoecology of the Fossil Vertebrates from the Central Narmadā Valley, Dist. Narsinghpur, Madhya Pradesh, India," which was undertaken under the overall supervision of the author, formerly of the Deccan College, Pune. The project has contributed immensely to an understanding of the geoarchaeology and man-land relationship in the Central Narmadā Valley in a number of ways (see concluding remarks). Because of the importance of certain aspects of study, a few results of our research under the IGRMS project have already been published, for example Microvertebrates, Rock Art research, etc. and these are presented here in updated form.

The successful implementation and completion of the project is due to the efforts of various agencies and individuals who were involved in the project in one way or the other. The author extends his grateful thanks to Dr K.K. Chakravarty, IAS (Member-Secretary and Trustee, Indira Gandhi National Centre for the Arts, New Delhi), for evincing keen interest in this work. Discussions with him from time to time have been very meaningful, illuminating and educative, and the author has benefited tremendously by continuous interaction with him, particularly on the broader perspective of museum movement and its role in environment and ecology, and various aspects of river valley cultures. It has been a great experience working with him on this and other projects. Dr Chakravarty offered all co-operation and help at every stage for the author's research, whether academic or administrative.

The then Acting Director, Shri Vikas Bhat, deserves grateful thanks for sanctioning a generous financial grant, and for his support and co-operation at every stage during the implementation of this project. Sri

D.K. Jain (IGRMS) was a member of the team and his co-operation and technical skills contributed greatly to the success of the project.

Thanks are also due to the authorities of the Deccan College, Pune, for providing various facilities from time to time and for granting duty leave in connection with the project. Being academically associated with Dr S.N. Rajaguru, the former Joint Director of the Deccan College, the author has benefited tremendously from numerous field trips and discussions with him. This association will remain a cherished memory for all time to come.

The author offers his grateful thanks to the authorities of the Archaeological Survey of India, M.P. State Dept. of Archaeology and Museums, Collector, Dy. Collector and Superintendent of Police of Narsinghpur Dist. for permission to conduct the excavations at various sites in M.P., especially in Narsinghpur Dist. and for their overall co-operation. Dr S.K. Acharya, Shri P.K. Basu, Shri S. Biswas, Shri Akhilesh Choudhary and Shri Sibdas Ghosh (all formerly of Geological Survey of India) gave many useful suggestions regarding the Narmadā geology for which the author is thankful.

A number of experts and students were associated at various stages during the execution of the project. These included (in the first field season) Dr Vijay Sathe (Research Assistant), Dr Rajeev Patnaik (who studied the micromammals collected from Devakachar), Ms Seema Pawankar (who worked as a Research Assistant on this project), Dr Kiran Kaul (a pedologist), postgraduate students Usha Gadre, Nayantara Shinde and Jyotsna Mauriya. Other staff included Chandrakant, Bhonsale and Laxman Singh. In the second field season, participants included Derek (from U.K.), Samantha Bhiladwala, Poornima Kirloskar, Jubal Fernandes, Sanatomba, Arun Kumar Singha and Krishna. Shri Rakesh Bhatt and Shri S.K. Pandey accompanied the author on a few field trips around Shahganj, Hathnorā, Kiriāghāṭ, Adamgarh and Bhīmbeṭkā in the third field season. All members of the team performed their duties extremely well, whether technical or logistic.

At the Deccan College, technical help was provided by Sunil Jadhav, Karanjikar, Vaman Gavde and Ravi. At the IGRMS, K. Seshadri, Ravi Shrivas, Anand Tehanguria, Praveen Narad and Dipak Chaudhury provided the technical help. Vijay Mohan, Shubhendra Kumar Singh, Joga Singh Rawat and others were happily ready to extend a helping hand whenever needed. Thanks are due to all of them and to the administrative staff of both IGRMS and Deccan College for their co-operation and help during the implementation of the project.

The author would like to extend thanks to his colleagues at Deccan College, Dr (Ms) Anupama Kshirsagar for analysing soil samples from the excavated profiles in the Devakachar area, Dr (Ms) Arati Deshpande-Mukerjee for studying the molluscan shells and Dr Rajeev Patnaik (from Panjab University, Chandigarh) for studying the microvertebrates discovered during the project research. All these contributions are compiled in the form of appendices attached with this book. Shri Sanjay Kumar (now late) helped the author in the study of some of the faunal material collected during the excavation and the author is grateful to him.

It gives the author great pleasure to acknowledge the help and co-operation of Shri S.B. Ota, formerly Superintending Archaeologist, Archaeological Survey of India, Bhopal (with an additional charge of the Director IGRMS) and Prof. K.K. Basa (present Director, IGRMS) for their keen interest in this work and help in seeing the book through the press. The author is also grateful to Dr. Narayan Vyas, formerly Deputy Superintending Archaeologist, ASI Bhopal, for his help and co-operation in the author's study of rock art. Late Prof. Shankar Tewari and his son Dr Anurag Tewari visited the excavation site in the Narmadā, the former enriched the author's knowledge regarding the geographical situation of the region and river behaviour of the Narmadā. Thanks are due to Sri S.B. Ota and Dr Sheila Mishra for their suggestions. Ms Bindiya Chopra and Sri P.K. Badam also deserve my thanks for their kind help.

There are many individuals in and around Narsinghpur and Devakachar but for whose help it would not have been possible to complete this work as successfully as it has been done. Special mention may be made of Rakesh Nema, his father (Sethi ji), Capt. Nema and several friends for their logistic support, Than Singh, Sri Goel and the Patel of Umeria for their help and co-operation from time to time. About twenty helpers worked in these or earlier field trips and shared with the author the many ups and downs of field research. They were just like family members and the author recalls the association of the villagers of the entire area with feelings of nostalgia. This work boosted the self-respect and sense of identity of the villagers, and deepened the mutual bonds of friendship.

The author's wife and children have been very helpful and co-operative in his academic ventures, through thick and thin, which kept him away from them for long spells of time. Heartfelt thanks are due to them for their understanding, patience and forbearance. It is a great pleasure to dedicate this book to them.

The author gratefully remembers the prehistoric animals in the Central Narmadā Valley who were transformed from biosphere to lithosphere, and continue to provide clues to the faunal distribution pattern and ecological history for the archaeologists, geologists and palaeontologists. It has been a great experience working with these mute spectators to the relentless passage of time.

Finally, the author is thankful to Shri Susheel K. Mittal of M/s. D.K. Printworld (P) Ltd. New Delhi, for bringing out this book meticulously in the present form and in a record time.

G. L. Badam

Contents

Key to Transliteration

VOWELS

अ *a*	आ *ā*	इ *i*	ई *ī*	उ *u*	ऊ *ū*
(but)	(palm)	(it)	(beet)	(put)	(pool)
ऋ *ṛ*	ए *e*	ऐ *ai*	ओ *o*	औ *au*	
(rhythm)	(play)	(air)	(toe)	(loud)	

CONSONANTS

Guttural	क *ka*	ख* *kha*	ग *ga*	घ *gha*	ङ *ṅa*
	(skate)	(blockhead)	(gate)	(ghost)	(sing)
Palatal	च *ca*	छ* *cha*	ज *ja*	झ *jha*	ञ *ña*
	(chunk)	(catch him)	(john)	(hedgehog)	(bunch)
Cerebral	ट *ṭa*	ठ* *ṭha*	ड *ḍa*	ढ* *ḍha*	ण* *ṇa*
	(start)	(anthill)	(dart)	(godhead)	(under)
Dental	त *ta*	थ *tha*	द *da*	ध* *dha*	न *na*
	(path)	(thunder)	(that)	(breathe)	(numb)
Labial	प *pa*	फ* *pha*	ब *ba*	भ *bha*	म *ma*
	(spin)	(philosophy)	(bin)	(abhor)	(much)
Semi-	य *ya*	र *ra*	ल *la*	व *va*	
vowels	(young)	(drama)	(luck)	(vile)	
Sibilants	श *śa*	ष *ṣa*	स *sa*	ह *ha*	
	(shove)	(bushel)	(so)	(hum)	
Others	क्ष *kṣa*	त्र *tra*	ज्ञ *jña*	ळ* *ḷ*	ॠ* *ṝ*
(*kṣatriya*)	(*triśūla*)	(*jñānī*)	(play)		

अं (ं) *ṁ anusvāra* (nasalisation of preceding vowel) like *saṁskṛti*

अः *visarga* = *ḥ* (aspiration of preceding vowel) like (*prātaḥ*)

ऽ *Avagraha* consonant #'consonant (like:- *ime 'vasthitā*)

Anusvāra at the end of a line is presented by m (म्) and not ṁ

HINDI LETTERS (extras)

ँ = ñ (*candrabindu*) ं = ṅ (*anusvāra*) ड़ = *ṛa* ढ़ = *ṛha*

* No exact English equivalents for these letters.

List of Plates

Chapter-3

Pl. **3.14 :** An *in situ* complete rib of a mammal in the section near Balaghāṭ.

Pl. **3.15 :** A skull of *Bos namadicus* excavated from a silty horizon at Luknu on way to Umeria, north of Devakachar. The horncores are complete. However, the front portion of the skull, which was buried in the soil, was damaged.

Pl. **3.16 :** A complete right horncore of *Bos namadicus*. The left horncore is buried in the gravelly horizon.

Pl. **3.17 :** A horncore from the top of clayey horizon in one of the gullies northwest of the pilot trench near Gurāriāghāṭ.

Pl. **3.18 :** A view of Sūkhā Nālā near Devakachar where a mandibular symphysis of *Elephas* was excavated from the base of the section. A non *in situ* ash deposit was also found in the gravel in the top portion of the section.

Pl. **3.19 :** Close up of the mandibular symphysis of *Elephas* (the symphyseal region got damaged during excavation).

Pl. **3.20 :** A trial trench south of the pilot trench towards the Sher-Umer; ten such trial trenches were dug to ascertain the lithological characteristics and thickness.

Pl. **3.21 :** An excavated trench which yielded horncores of *Bubalus* sp.

Pl. **3.22 :** Horncores of *Bubalus* sp. from 3.21 excavated trench.

Pl. **3.23 :** A section on the left bank of Narmadā near Barmanghāṭ below the PWD Rest House where an almost complete burial of a huge buffalo was excavated.

Pl. **3.24 :** A lower portion of the dolerite dyke at Barmanghāṭ where a large number of Stone Age tools were collected in addition to some fossils, preserved as lag deposits after the fluvial transport.

Pl. **3.25 :** A thick section of the alluvium near Jhānsighāṭ (on way to Jabalpur) where a lot of faunal and cultural material was found.

Pl. **3.26 :** A general view of the Narmadā at Hathnorā (northeast of Hoshangabad). The bouldery gravel bed (sheet deposit) yielded a partial cranium of *Homo erectus narmadiensis* in 1984. It is a GSI find.

List of Figures

Chapters

Appendix-I

Hydrobiidae indet. (VPL/RP-M5); **f.** Cf. *Gyraulus* sp (VPL/RP-M6); **g.** *Pisididae* indet. (VPL/RP-M7); **h** & **i.** isolated tooth and scale of *Cyprinidae* indet (VPL/RP-M8&9). Bars represent 5mm.

fig. 4. Dental terminology of murid first upper and lower molars (after Jacobs, 1978).

fig. 5. *Millardia* cf. *M. meltada.* **a.** occlusal; **b.** lingual and **c.** labial views of maxilla (PLDC-M-1); **d.** occlusal; **e.** lingual and **f.** labial views of mandible (PLDC-M-5). Bar represents 1 mm.

fig. 6. *Bandicota* cf. *B. bengalensis.* **a.** occlusal; **b.** lingual and **c.** labial views of mandible (PLDC-M-11), bar represents 2 mm. *Tatera el T. indica;* **d.** occlusal view of maxilla (PLDC-G-1), bar represents 1 mm. *Gerbillus* sp.; **e.** occlusal; **f.** lingual; and **g.** labial view of maxilla with M (PLDC-G-2). Bar represents 2 mm.

fig. 7 **A.** Longitudinal section of upper incisor of *Tatera* cf. *T. indica;* B-E, *Bandicota* cf. *B. bengalensis;* **B.** Upper portion of longitudinal section of upper incisor; **C.** lower part in detail; **D.** upper portion of the transverse section of upper incisor; **E.** longitudinal section of the lower incisor. Bars represent 10 mm.

fig. 8. Scatter diagram showing size variation within Nesokia and Bandicota extant specimens, *Bandicota indica, Bandicota bengalensis,* and *Nesokia indica* and fossil specimen, *Bandicota sivalensis* (from Patnaik 1991).

fig. 9. Gerbillid dental terminology, a, M and b. M1 (after Tong, 1986).

fig. 10. Soricidae indet. **a.** labial view of the mandible (PLDC-1-1). *Corcodylus* cf. *C. palustris.* **b.** anterior tooth (PLDC-R-1) and **c.** posterior tooth (PLDC-R-2). *Anura indet.* **d.** linguo-occlusal view of the maxilla (PLDC-A-1). *Cyprinus* cf. *carpio communis.* **e.** pharangeal arch with one tooth (PLDC-P-1). Bars represent 1 mm.

fig. 11. Scatter diagram showing teeth dimensions of *Crocodylus palustris* (o-maxillary and x-mandibular teeth) and *Crocodylus* cf. *C. palustris* (o-mandibular teeth). **a.** ZSM 3411912; **b.** ZSM 51711911 and **c.** ZSM 23 110 (modified from Patnaik and Schleich 1993).

fig. 12. **a.** Skull cap of *Bandicota* cf. *B. bengalensis* (PLDC-M-10); **b** & **c.** Femora fragments of rodents; **d.** Scapula of rodent; **e.** and **f.** lower jaws of *Millardia* cf. *M. meltada* Bar represents 2 mm.

fig. 13. Hypothetical reconstruction showing the various stages of fossil accumulation and sedimentation.

fig. 14. Hypothetical reconstruction of the various communities present and deposition of Devakachār Formation in the Narmadā Valley.

Appendix-II

fig. 1: Fresh water shells from the Central Narmadā Valley.

Appendix-IV

fig. 1 Location of Bhīmbeṭkā, District Raisen, M.P.

fig. 2 Proposed Palaeoecological Model at Bhīmbeṭkā.

1

Introduction

The rich heritage of the Narmadā is evident from the numerous places of historical and religious significance that it cradles, the eulogies in ancient literature that goes to its credit and the important status it holds in the hearts of people living not only on its banks but also in other parts of India. The sacredness of the river has not diminished through ages; the Narmadā Parikramā* (procession around the river), prevalent in the ancient times, is still being carried out as it is believed that this brings good luck to a person. The importance of Narmadā is so great that it is often referred to as "Narmadā Maiyā," equating it with one's mother.

Narmadā has a number of *ghāṭ*s of historical and religious significance on its banks, of which the important ones are: (i) Śāṇḍilya *ghāṭ*, where saint Śāṇḍilya meditated, (ii) Seṭhānī *ghāṭ*, where Māgha Pūrṇamī *melā* is held annually in the month of January, and (iii) Nemabar *ghāṭ*, where a Siddeśvari temple is located.

Three kilometres south of Hoshangabad Railway Station are located the famous Mesolithic rock shelters of Adamgarh, known for its rock art ever since it was first discovered by Raisahib Manoranjan Ghosh in 1919. Other rock art sites in the region are Tamia, Pañchamaṛhī, Śivni-Malwa, Bhīmbeṭkā, Sāñchī, Nāgorī, Raisen and Narsinhgaṛh. The important dynasties that reigned in valley were the Vākāṭaka, Nāga dynasty in the Māṇḍlā region and Kadamba dynasty in the Chindwāḍā region. Of the people living in the valley, an important tribe to mention here is the Kutch-Beliyas who are semi-nomadic, polygamous and used to practice hunting and gathering in the recent past but have now integrated to more settled way of subsistence.

The Narmadā Valley has attracted the attention of anthropologists, geologists and pre-historians for more than a century and a half. However,

* The author was leader of a team of various scholars conducting Narmadā Parikramā, organized by IGRMS, Bhopal. Results of this Parikramā will be published shortly.

it is not necessary here to mention the works of all scholars as most of these are published. Therefore, a brief reference to only some of the important ones is given.

Previous Work and Present Research

Captain Sleeman was the first to discover mammalian fossils from the area around Jabalpur in 1830, an account of which appeared later in Princep (1832). Subsequently, Princep (1833) and Splisbury (1837, 1844) made rich collection of fossils from various sites in the Narmadā. However, the first systematic study from litho- and biostratigraphic points of view was done by Theobald (1860) who divided these beds into two groups, Lower and Upper and assigned them a Pliocene age. Subsequently, Medlicott (1860), Blanford (1869) and Oldham (1871, 1893) conducted research on various aspects of the Pleistocene history of the Narmadā. Lydekker (1880, 1882, 1884), who was simultaneously working on the Śiwālik deposits of NW India as well, made remarkable studies on the fauna of both the regions. On the basis of the find of a handaxe (collected by C.A. Hacket from near Bhutra, north of Gadarwara) Medlicott (1873) dated the beds as late Pleistocene. Pilgrim (1905) after studying the fossil fauna from the Godāvarī and Narmadā river valleys, suggested that the Narmadā deposits must be considered as of Middle Pleistocene age. de Terra and Paterson's (1939) multidisciplinary investigations gave fresh impetus to the study of Pleistocene in the Narmadā Valley. They studied the region between Hoshangabad and Narsinghpur in detail and tried to correlate these deposits with those of the Śiwāliks of northern India. In addition, they attempted to correlate the Narmadā Stone Age industries with the Sohan culture of NW India. According to them both the Lower and Upper groups of the Narmadā alluvium are characterized by the presence of Middle Pleistocene faunal elements. Sankalia's work (1974) around Maheśvar contributed further to our understanding of the stratigraphical and cultural sequence of the Narmadā deposits. According to Khatri (1961, 1963, 1966) who discovered several palaeontological and implementiferous sites in the valley, there occurred no faunastic change in both the Lower and Upper groups of the Narmadā. Sen and Ghosh (1963) suggested a Middle Pleistocene age for the Lower Narmadā Group and a probable Upper Pleistocene age for the Upper Narmadā Group. They further opined that there must be a faunal difference between the two groups. Supekar (1968) discovered more Palaeolithic sites and fossil localities and ascribed an early Upper Pleistocene age to the Lower Palaeolithic industry.

About three decades ago, Badam (1979a) undertook detailed palaeontological studies of the material collected by him from various localities between Bheḍaghāṭ and Hoshangabad and tried to build up a relative timescale for the Palaeolithic cultures on the basis of fauna. Along with his collaborators and students, Badam published a lot of literature bearing upon various aspects of litho- and biostratigraphy, cultural materials, palaeoecology and taphonomy, which have contributed immensely to a better understanding of the chronology and palaeo-environment of the area as a whole (Badam, 1979b, 1982; Joshi, et al., 1978; Badam and Salahuddin, 1982; Salahuddin and Badam, 1983; Badam, et al., 1986; Badam and Grigson, (1990). Biswas and Dassarma (1981) after a detailed investigation of various sites in the valley collected a lot of faunal material the study of which threw new light on the chrono- and lithostratigraphy of the area as a whole.

In addition, reports of more discoveries of fossils from the valley including ash beds and their impact on litho- and chronostratigraphy have briefly appeared in *Indian Archaeology — A Review*, *News Bulletin of the Society of Vertebrate Paleontology* (USA), *East Asian Tertiary/Quaternary Newsletter* (Hongkong) and *Geological Survey of India News*.

Researches on the Narmadā Valley evoked further interest from various scholars, both Indian and foreign, with the unique and the only discovery, so far, of *Homo erectus narmadiensis* by Sonakia (1984, 1985) from Hathnorā (22° 52' N: 77° 53' E), about 40 km north-east of Hoshangabad.* This discovery has filled to some extent the gaps in our understanding of the distribution and evolution of the hominid fossils in South Asia as a whole. Though there has been a difference of opinion about its *in situ* position and chronological framework, this has been the first authentic discovery of its kind in India. Scholars who subsequently worked on various aspects of the hominid stratigraphy include Badam et al. (1986), Salahuddin et al. (1986-87), Badam (1989), Kennedy et al. (1991) to mention a few. They came out with interesting results regarding the analysis of the cranium, its geological and archaeological association and its chronology.

Another notable discovery which has an important bearing on the chronology and palaeo-environment of the Central Narmadā Valley is that of microvertebrates from Gurāriāghāṭ near Devakachar in Narsinghpur Dist. (Patnaik et al. 1995; Patnaik, 1995). This is the first discovery of microvertebrates from Peninsular India with the exception

* In recent years more discoveries of the hominid fossils have been made from Hathnorā by A.R. Sankhyan, 2005.

of Kurnool Caves in Andhra Pradesh, and indicates an early emergence of the modern rodents of the Indian subcontinent. The collection which includes micromammals, fish, amphibians and reptiles came from a fresh water mollusc-bearing pebbly horizon exposed on the right bank of the tributary Sher-Umer near Devakachar (23° 23' N: 79° 07' E) in Narsinghpur Dist.

It will not be out of place to mention here that not only has the Narmadā Valley yielded the relics of human cultures right from Stone Ages to the Medieval times, there are numerous rock shelter sites very close to the Narmadā as mentioned above, which have preserved artistic representation of the Early Man from Palaeolithic to Historical periods. One of these shelters has been excavated by the author in the IGRMS campus, Bhopal. Further, a preliminary report of the ecological aspects of Bhīmbeṭkā rock shelters is attached with this book in the form of an appendix for a better understanding of the ecological history of the entire region.

All the above discoveries made in the last century-and-a-half on diverse aspects of the Narmadā Valley have established the significance of the valley in relation to geology, archaeology, palaeontology and history. The microvertebrates collected under the project have been published earlier, but an updated version is included in this book for an overall understanding of the material and its importance (see Appendix I by Rajeev Patnaik). Though molluscan shells have been found in great numbers from the Narmadā Valley, no attempt has so far been made to study the material in detail from morphological point of view. For the first time, a preliminary analysis of some of the molluscan shells which have an important bearing on the chronology and environment of deposition of the Narmadā sediments has been done by Ms Arati Deshpande-Mukerjee (see Appendix II). Even though chemical analysis of soil and bone samples collected randomly from the Narmadā Valley has been published earlier, the analysis of the soil samples collected from excavated profiles in Devakachar and Umeriā is attempted in this work to have a better understanding about the mode of deposition of the Narmadā sediments (see Appendix III by Anupama Kshirsagar). Appendix IV by G.L. Badam amd V.G. Sathe summarizes the palaeoecological background of Bhimbeṭkā rock shelters showing the close link with the palaeoecology of the Narmadā Valley. This is based on published work.

Stone Age tools collected from excavated profiles especially at Gurāriāghāṭ in Devakachar area have also been studied with special emphasis on the typology and techniques though this work is not included here in detail. Thus on the whole a multidisciplinary approach has been attempted on the study of part of the Central Narmadā Valley which is expected to bring to light several important aspects of this interesting river valley.

2

General Account and Geology

THE Narmadā, one of the major rivers of India, flows across the middle of the Indian subcontinent and forms a traditional dividing line between the north and the south. The river originates from the plateau of Amarkaṅṭak (22° 40' N: 81° 46' E) in the Maikala range of eastern Madhya Pradesh and flows for a distance of around 1300 km through Madhya Pradesh and Gujarat via a small stretch in Maharashtra before debouching in the Gulf of Cambay (*fig*. 2.1). During its course the river flows through Deccan Traps (of Cretaceous period), crystalline dolomitic limestone (Dharwar System), fertile plain flanked by massive sandstone and quartzite formations of Vindhyas and Satpuṛās and finally through alluvial country. Its major tributaries from the southern flank are the Sher, the Umer, the Varurewa, the Shakkar, the Dudhi and the Tawa. The Hiran and the Sindhar are amongst the few tributaries which join the Narmadā from the northern side.

At Bheḍaghāṭ (23° 08' N: 79° 47' E) near Jabalpur, the Narmadā flows over a fall of about 9 m (as a result of tectonic disturbances during late Tertiary and early Pleistocene periods) and thereafter flows in the form of a saucer shaped basin between Bheḍaghāṭ and Haṇḍiā near Hoshangabad (22° 45' N: 77° 45' E) (*figs*. 2.2 & 2.3). This basin preserves about 150 m of Cenozoic sediments, the upper 50 m or so of which have yielded one of the richest Pleistocene faunal assemblages in India. In addition to faunal elements, geological and cultural units are also well preserved in this trough.

Since the present work pertains mainly to the Narsinghpur District, a general account of climate and biota of the district is given here:

Narsinghpur District

CLIMATE

The Narsinghpur district has tropical type of climate with hot summers (March-June), monsoonal rain (June-September), post-monsoonal season (October-November) and cold winters (November-February), the month of July being the most wet. During the monsoons, the weather in the district

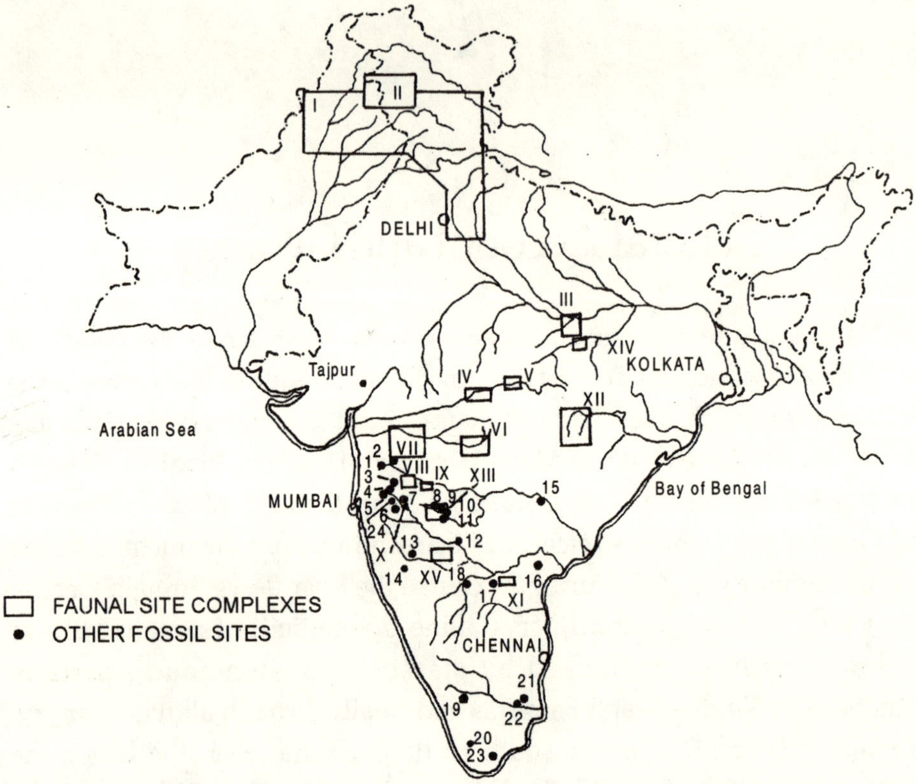

I	Śiwālik hills — Punjab-Haryana-Jammu-Himachal Pradesh	1	Nasik	17	Betamcherla
		2	Nandur	18	Nittur
II	Karewa formation — Kashmir	3	Kalamb	19	Coimbatore
III	River Belan — Allahabad dist.	4	Bori	20	Sayamalai
IV	River Narmadā — Hoshangabad	5	Chandoli	21	Ariyalur
V	River Narmadā — Narsinghpur	6	Bhivari	22	Maruvattoor
VI	River Pūrnā — Akolā	7	Inamgaon	23	Ayyanidipu
VII	River Tāpī — Dhūliā dist.	8	Dhanegaon	24	Walki
VIII	River Pravarā — Ahmednagar	9	Tadula		
IX	River Godāvarī — Aurangabad	10	Wangadari		
X	River Ghoḍ — Pune dist.	11	Ganjur		
XI	Kurnool caves — Kurnool dist.	12	Hagargundi		
XII	River Mahānadī — Raipur dist.	13	Sirguppi		
XIII	River Manjra — Bhir and Latur dist.	14	Chikdauli		
XIV	River Son — Sidhi dist.	15	Servaipet		
XV	Hunsgi Baichabal — Gulbarga dist.	16	Venukonda		

fig. 2.1: Important Pleistocene fossil bearing sites in India and the position of Narmadā river.

becomes pleasant due to showers of rain and the relative humidity goes up to 70 per cent. During the other seasons, the relative humidity is comparatively low sometimes reaching down to 20 per cent in the summers. During the summers, the maximum mean daily temperature

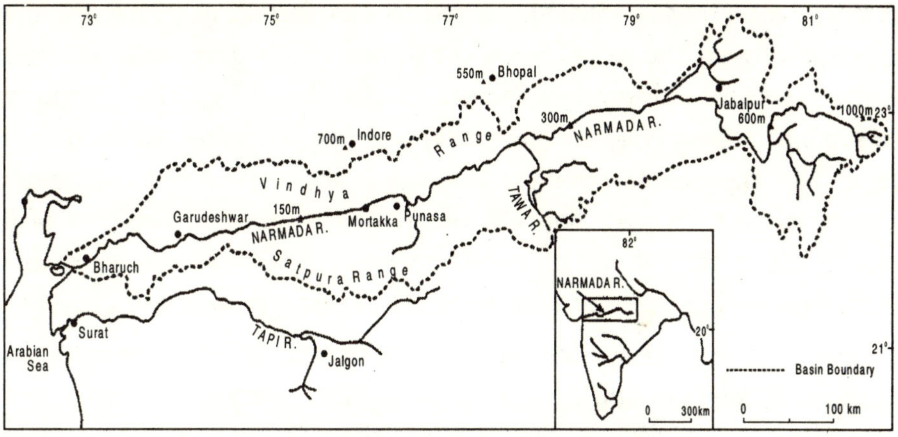

fig. 2.2: The Narmadā Basin.

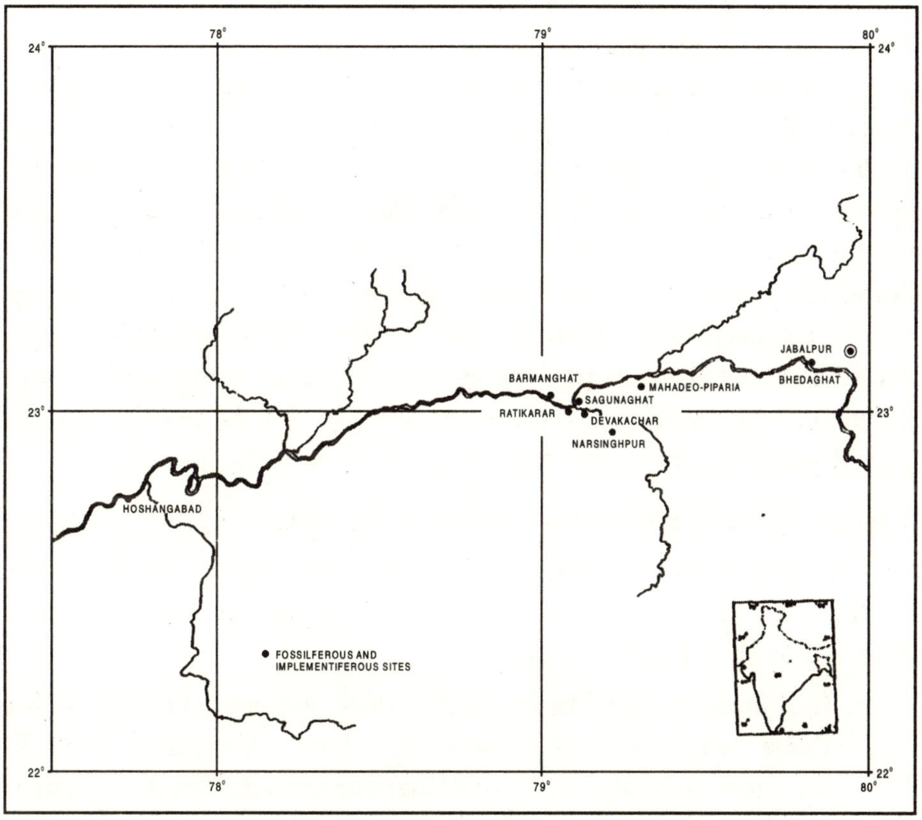

fig. 2.3: Central Narmadā Valley : The Area of research.

exceeds 40°C sometimes attaining even a maximum of 47°C and the minimum mean daily temperature remains at about 25-26°C. The winters are cold and dry with the maximum and minimum mean daily temperatures being about 25°C and 9°C respectively.

FLORA AND FAUNA

The vegetation cover in the district is of tropical dry type with a few patches of tropical moist type which constitutes mixed deciduous forest. ***Sāl*** (*Shorea* *robusta*), teak (*Tectona grandis*), mixed forests and shrubs form the main vegetation cover. The mixed forest constitutes of ***sāj*** (*Terminalia tomentosa*), tendu (*Diospyros melanoxyion*), ***mahuā*** (*Meduca longifolia*), ***bīja*** (*Pterocarpus marsupium*), ***jāmun*** (*Eugenia jambolana*) and ***ber*** (*Zizyphus jujuba*) to name some. The important shrubs are ***karondā*** (*Carissa spinarum*) and bamboo (*Dendrocalamus strictus*). The ***sāl*** forest covers about 80 per cent of the area mostly in the flat and undulating terrain. Apart from these, a variety of climbers and grasses grow in the forest as undergrowth and grasslands.

The present fauna of the region comprises a variety of mammalian, reptilian, amphibian and piscean species. The common mammals found here are bear (*Malursus ursinus*), monkey (*Macaca mulata*), ***laṅgūr*** (*Semnopithecus entellus*), ***nīlgāi*** (*Boselephus tragocamelus*), sambhar (*Cervus unicolor*), ***bārāsiṅghā*** (*Cervus duvauceli*), blackbuck (*Antilope cervicapra*), gazella (*Gazella gazella bennetti*), ***chital*** (*Axis axis*) and wild boar (*Sus scrofa cristatus*). Apart from these, jackal, fox and wolf are also found in the jungles. Of the reptiles the important ones to mention here are short snouted crocodile (*Crocodylus palustris*), cobra (*Naja tripudions*), crait (*Bungarus pasciatus*), ***ajagar*** (*Python* spp.) and turtle. Elephants which were common till the 16th century are absent from this region now. Tigers and leopards are also fast decreasing in number due to deforestation and other human activities. Gaur occasionally comes to the hills of south-west around Mohapani and Dudhi rivers. ***Chinkara*** and blackbucks, two endangered species, are commonly seen in the southern part of the district.

Geology of the Central Narmadā Valley (*fig.* 2.4)

A continuous geological sequence from the pre-Cambrian to the present is well preserved in the valley. The hard rock geology of the valley can be divided into six categories:

(i) ***Bijawar Series:*** The cherty limestone and breccia of the Chawar Patha tract to the north of Narmadā belong to this series. The rocks are often co-related with the Dharwar system. Phyllites, schists, quartzites and limestones occur south of Narmadā. Rocks exposed around Jabalpur are metamorphosed, folded and faulted.

(ii) ***Granite and Gneiss:*** Rocks made of quartz, microline, biotite and muscovite occur ar various places. Schists are impregnated with gneisses.

(iii) ***Vindhyan System:*** It is of late pre-Cambrian age and is made up of hard, red sandstones bearing ripple marks.

(iv) ***Goṇḍwānā System:*** It is made up of about 3000 m of fresh water deposit distinguishable into Upper and Lower Goṇḍwānā.

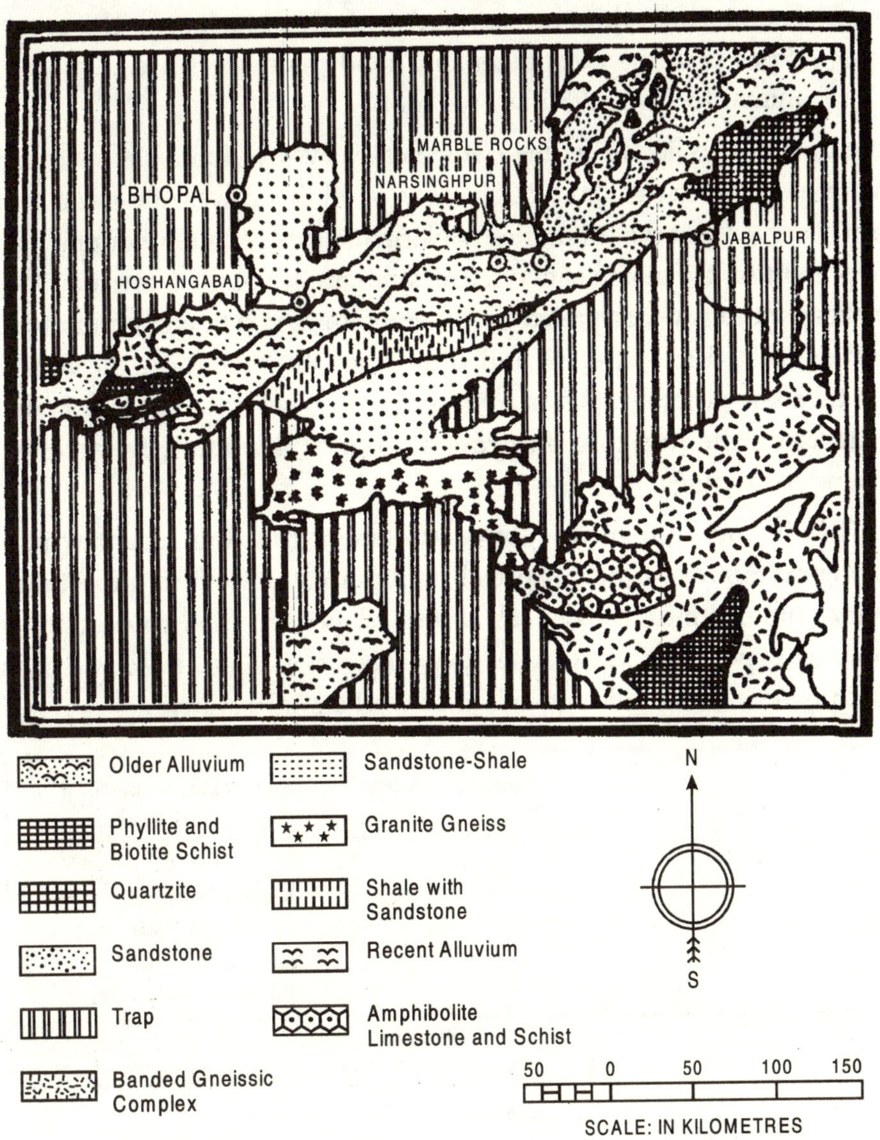

fig. 2.4: Geological Map of the Central Narmadā Valley.

The former group is characterized by sandstones alternating with clays and conglomerates and the latter by green boulder clays with sandy lenticules and coarse sandstone.

(v) ***Deccan Traps***: A large portion of the Central Narmadā Valley is covered by the basaltic Deccan Trap which is not uniform in thickness. At many places Trap is tilted and faulted.

(vi) ***Alluvium***: The thickness of the alluvium, composed of gravels, sands, silts, and clays varies from place to place. Jabalpur rock forms a thin sandy soil, Dewās conglomerates produce red clay filled with pebbles. Bogra conglomerates give rise to very little soil while the uplands of Satpuṛā are covered with thick sandy soil.

The metamorphics are covered with thin black soil. The sections exposed in the river beds and flood plain areas are very rich in animal fossils and Stone Age tools.

Bore hole data collected by the Geological Survey of India shows that the Quaternary deposit is about 50 m thick on both sides of the river. It has been observed that the tributaries of Narmadā lack certain lithounits which are present on the main channels. The stratigraphy of the Narmadā Valley has been variously interpreted by different authors. Different names have been suggested for different and sometimes the same horizon. It is felt that till some generalization is achieved, a fixed set of terms may be used to describe the horizons in order to avoid unwanted fallacies and misunderstandings.

It is in the light of the above statement that the recent work of Geological Survey of India (Special Publication 46, 1997) is important as, besides other things, a fixed set of terms has been used to describe the horizons exposed in various sections (Tewari & Bhai, 1997), a scheme which is now being followed by other workers in the area.

QUATERNARY	Recent/Holocene	Ramnagar Formation Bauras Formation
	Pleistocene	Hirdepur Formation Baneta Formation Sūrajkuṇḍ Formation Dhansi Formation Pilikarar Formation
..... Unconformability		
PRE-QUATERNARY	L Eocene to Upper-Cretaceous	Deccan Basalts
	Cretaceous	Lametas
	Upper Jurassic to Permo carboniferous	Goṇḍwānā Super Group
	Proterozoic	Vindhyan Super Group Bijawar Super Group Mahākośala Group
	Achaeans	Gneisses and Granite

3

Lithostratigraphy of the Alluvium

THE Narmadā Valley is flanked on the northern side by the mountain ranges of the Vindhyas and on the southern side by those of the Satpuṛās. Geologically, the Archaeans are the basalmost complex followed by Bijawar Series, Vindhyans, Goṇḍwānās, Deccan Traps, Laterite beds and lastly the Alluvium. Theobald (1860) was the first to suggest a chronological sequence of the Narmadā beds as follows and dated these beds as Pliocene.

Upper Group	Black Cotton Soil Pale yellowish brownish alluvium Irregular beds of sands and conglomerates
Lower Group	Red yellow clay Boulder Conglomerate

de Terra and Paterson (1939) established the following stratigraphical sequence of the Narmadā alluvium, suggesting a Middle Pleistocene age for the beds.

Younger alluvium	Cotton Soil Group	Black Cotton soil Basal gravel
	-----Disconformity-----	
	Upper Group	Upper pink concretionary silt Upper gravel and sand
Older alluvium	-----Disconformity-----	
	Lower Group	Lower red concretionary clay Basal Conglomerate
	-----Disconformity-----	
	Laterite gravel and Laterite silt	

Khatri (1966) considered the beds as of Middle Pleistocene to Holocene age and gave a sequence as given below:

Yellow silt + Modern soil
Yellow cross-bedded sand
Cemented sandy gravel
Red clay/Boulder Conglomerate

Supekar (1968) suggested the following composite stratigraphy based on observation of several sites in the Narmadā Valley and also his excavations at Mahādeo-Pipariā (a site about 25 km north-east of Narsinghpur, on the left bank of the Narmadā), and assigned an early Upper Pleistocene age to the beds.

Black Cotton Soil	Microliths
Yellow brown concretionary silt with occasional fine sandy lenses	Upper Palaeolithic (poorly defined)
Sandy pebbly gravel with occasional cross-bedding	Middle Palaeolithic
Red brown concretionary silt	Middle/Lower Palaeolithic
Boulder Conglomerate	Lower Palaeolithic

Badam (1979a) considered the age of the Narmadā beds to range from late Middle Pleistocene to early Upper Pleistocene and gave the following stratigraphy of the alluvium:

Dark brown consolidated silt and gravel
Yellow brown calcreted silt, sand and gravel
Reddish brown calcareous silt
Cemented boulder gravel

Biswas and Dassarma (1981) assigned an Upper Pleistocene age to the Narmadā beds and gave the following sequence:

Black Cotton Soil

Grey silt

Pedocol layer

Grey sand with lime concretions

Pinkish clay and silt, lime concretions

Grit and sand

Boulder bed

Deep red clay

Deep red boulder bed

It can be seen from the various litho-successions given by the above workers that the lithounits are more or less similar with the difference that minute demarcations are suggested as based on the colour and geological units. However, most of the workers have broadly followed the scheme suggested by de Terra and Paterson (1939). Also, there are variations in lithology on the sections of the tributaries of the Narmadā and those on the main channels. The borehole data presented by Supekar (1968), in which gravel layers at intervals of 21.65 m, 29 m, 39 m, and 76.5 m below surface were tapped, suggested that the proximity of suitable lithological (e.g. Bogra

Conglomerates) and topographical (e.g. Vindhyan escarpment) features have been responsible for the strata building of the valley. This is further supported by the presence of cemented, bouldery gravel bed in the proximity of the Narmadā river and by fine sand and silt, as its homotaxis, in the tributaries of the river. Such variation in the lithological composition within the valley can be explained by the lateral migration of the river (Salahuddin et al., 1986-87). Therefore a comprehensive stratigraphy of the valley involves a detailed study of the subsurface alluvium.

Stratigraphical Sections

Though till date more than twenty sections on the main channels of the Narmadā and its tributaries have been excavated and studied in detail by the author and his collaborators, mention is made here of only those sites which were excavated and studied under the IGRMS project.

HATHNORĀ (*fig.* 3.1; *Pls.* 3.26 & 3.27)

This is one of the most important sections in the Narmadā Valley having yielded the fossil of a partial cranium of *Homo erectus narmadiensis*, the only hominid fossil found in India so far.* The site proper is situated on the northern bank of the Narmadā about 40 km north-east of Hoshangabad. Here a section of about 15 m is exposed at the site of discovery of the cranium. Granted that the lithological units are time transgressive, we have deviated from the traditional scheme of the Narmadā stratigraphy and divided the exposed alluvium into five lithounits. The basal unit of about 1.5 m thickness comprises greyish, cemented, cross-bedded sand and matrix supported pebbles at the top. Overlying this unit conformably is the cemented bouldery/pebbly gravel of 0.5 m thickness which has yielded the hominid cranium. In addition, fossil bovids, proboscideans and Chelonia have been collected along with late Acheulian tools from these gravels. This unit grades into a 5 m thick unconsolidated sandy/pebbly gravel having sub-rounded to rounded and unimbricated matrix supported pebbles/gravels. This is the most richly fossiliferous unit elsewhere in the Narmadā Valley, and has yielded Middle Palaeolithic and late Acheulian artefacts at Hathnorā. A pebbly/cobbly and silty clay unit of 2 m thickness overlies the previous unit and this has yielded Middle Palaeolithic artefacts. Capping it is the yellow silty clay of 4 m thickness which is in turn overlain by Black Cotton Soil.

---

* Other discoveries of fossil hominids from Hathnorā, e.g. clavicles have been reported by A.R. Sankhayan.

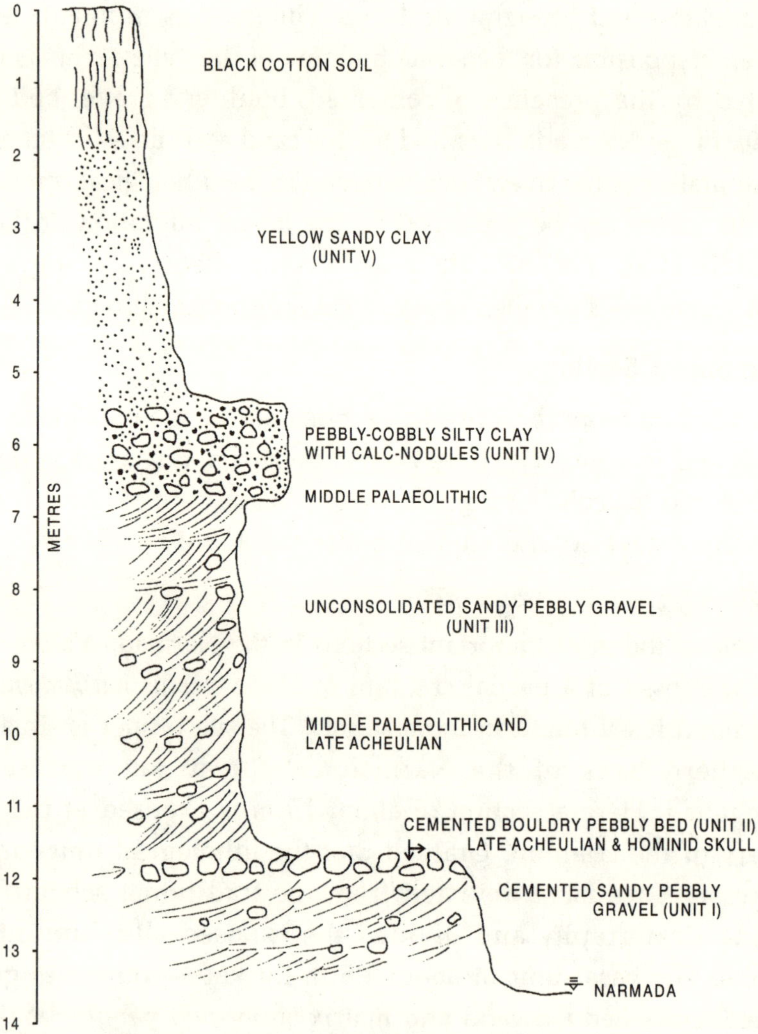

fig. 3.1: Lithostratigraphy at Hathnorā (The hominid-bearing section).

HOSHANGABAD

A few sections exposed in and around Hoshangabad include Kaṅkar Ghāṭ Circuit House section and Doṅgarwāṛā (3 km downstream from Kaṅkar Ghāṭ). A sandy pebbly gravel of about 1.5 m thickness forms the base of stratigraphical columns in these sections. The pebbles and cobbles are well rounded and made of sandstone, quartzite, basalt, chert, jasper and quartz. Capping this unit is 8 m thick yellow brown concretionary silt which is covered by Black Cotton Soil. In other sections the thickness of the beds varies and even the sequence of lithounits changes, e.g. instead of the basal sandy pebbly gravel some sections have concretionary silt at the base. But for these variations the general sequence of lithounits in and around Hoshangabad remains the same. A lot of faunal and cultural material was recovered from these sections.

SHAHGANJ

About 25 km north-east of Hoshangabad, this area exposes, on the upstream and downstream, well preserved sections of reddish brown concretionary silt and clay with intercalations of pebbly gravels in between. Important sections that were explored are Kusumkheḍā Ghāṭ (near Barkheṛā, 5 km away), Khaḍiā Ghāṭ (near Jayath), Dhansi, Sadar Nagar, Kiriāghāṭ, Sūraj Kuṇḍ, etc. and trial trenches were dug in some of these sections. Kiriāghāṭ turned out to be a richly fossiliferous site having yielded an almost complete skull of *Hexaprotodon palaeindicus* (*Pls.* 5.17 & 5.18) in addition to fossils of bovids, equids, turtles, etc. Cultural material is, however, scanty.

DEVAKACHĀR (*figs.* 3.2 to 3.6; *Pls.* 3.1 to 3.22)

This village situated about 10 km north north-west of Narsinghpur has exposed one of the best preserved cultural materials and abundance of

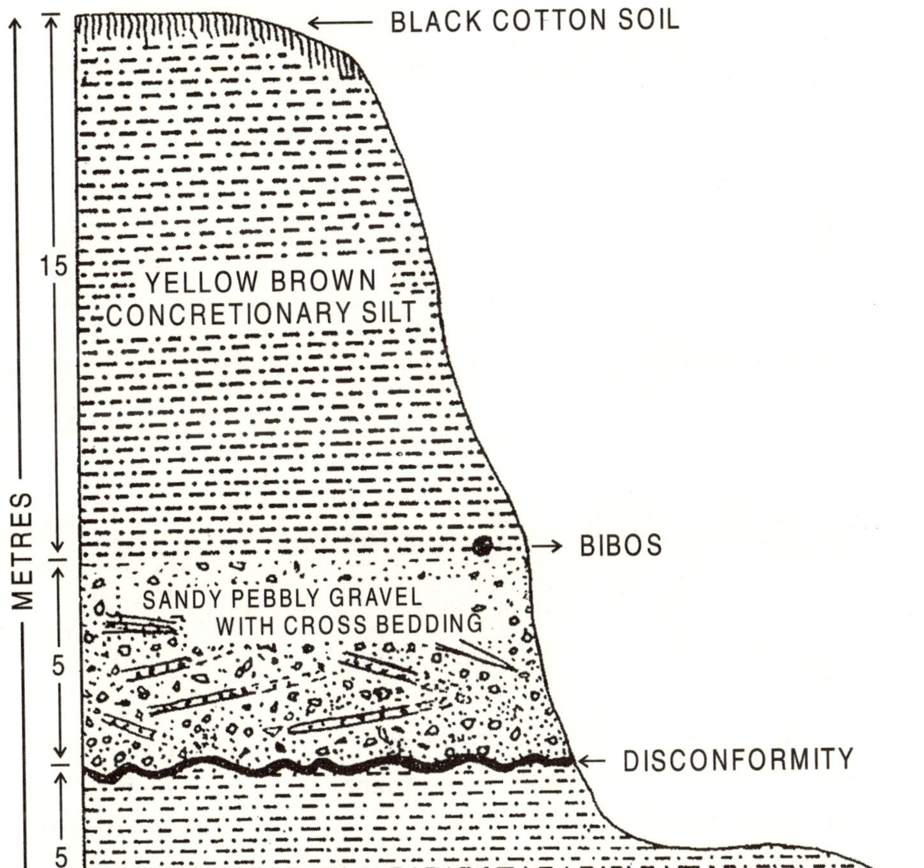

fig. 3.2: Section near Devakachār.

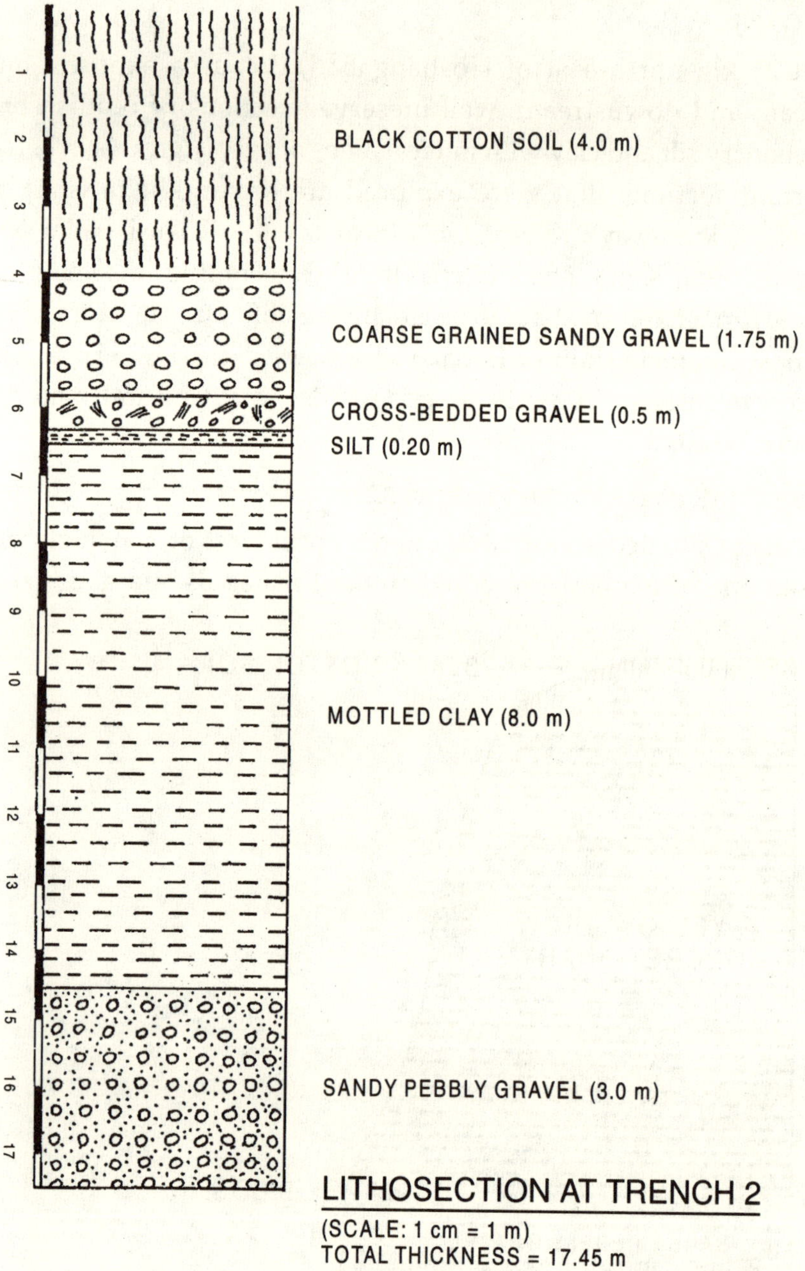

fig. 3.3: Lithosection at Gurāriā Nālā, Trench 2 (Devakachār Area).

faunal material on the right bank of the river Sher (one of the southern tributaries of the Narmadā) near its confluence with another tributary, the Umer. About 100 m downstream from Devakachār, a third tributary, the Varurewa meets the Sher-Umer and flows north for a few km before meeting the Narmadā proper flowing in an east-west alignment.

This entire area from a little upstream of Devakachār (Belghāṭ) to the confluence of the various streams (Sher-Umer, Varurewa) has been the

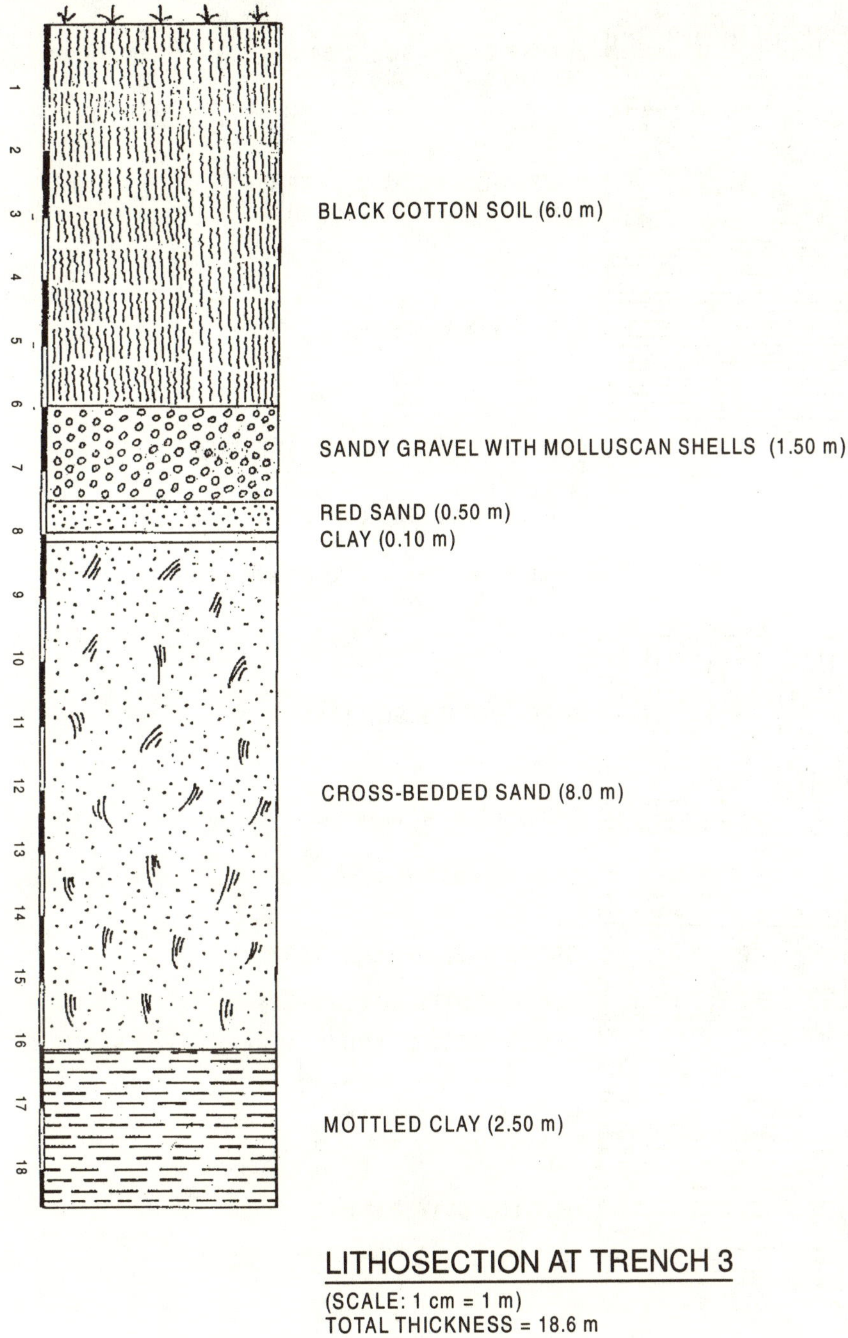

fig. 3.4: Lithosection at Trench 3 (Devakachār area).

scene of palaeontological activity since more than sixty years, because the horse-shoe meanders on the Sher-Umer and the sections exposed on the various streams mentioned above are one of the richest fossil-yielding sites in the Narmadā Valley. Also, away from the streams and located in the gullies are present numerous fossiliferous sections which have been explored/excavated from time to time.

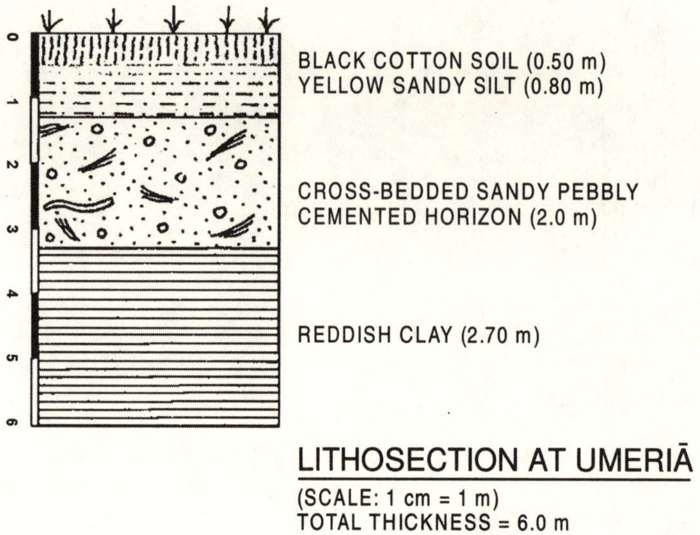

fig. 3.5: Lithosection at Umeriā (Devakachār area).

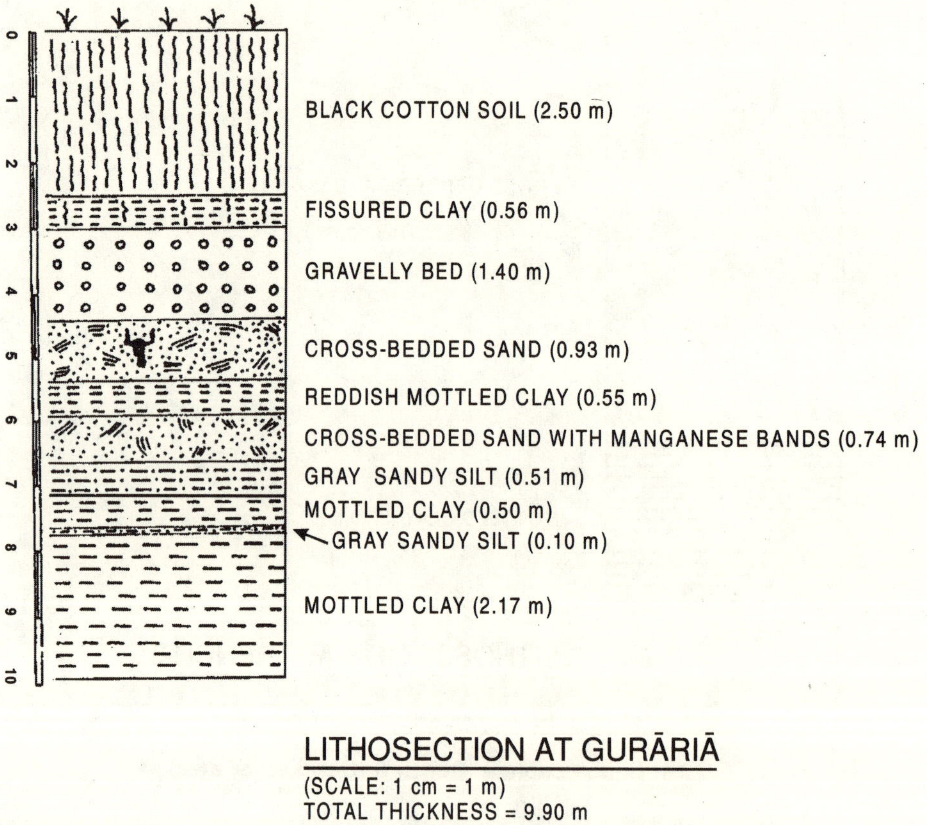

fig. 3.6: Lithosection at Gurariāghāṭ (Devakachār area).

Before working on the present project the author had explored vast areas and excavated a few sections in this region for a better understanding of the distribution pattern of fossils (Badam, 1979b). One former pupil of

Pl. 3.1: A view of the Sher-Umer Nadi from Devakachar, Narsinghpur Distt., flowing north towards the Narmadā.

Pl. 3.2: Top of the alluvium on the right bank of Sher-Umer.

Pl. 3.3: A view of the Gurāriāghāṭ section on the right bank of Sher-Umer which was treated as a pilot trench.

Pl. 3.4: The author discussing a point with late Prof. Shankar Tewari on the Gurāriāghāṭ section.

Pl. 3.5: Clay and sandstone beds at Gurāriāghāṭ section.
(the author with late Prof. Shankar Tewari)

Pl. 3.6: A view of the Gurāriāghāṭ section in which vertebrate fossils, molluscan shells, microvertebrates, Stone Age tools and a worked bone were recovered.

Pl. 3.7: The microvertebrate bearing gravelly horizon almost at the top of the section.

Pl. 3.8: A profile of the pilot trench showing stratigraphic details (clay and sandstone beds).

Pl. 3.9: Excavation of the skull of *Bos namadicus* in the gravelly silty section of the pilot trench. Note the broken horncore.

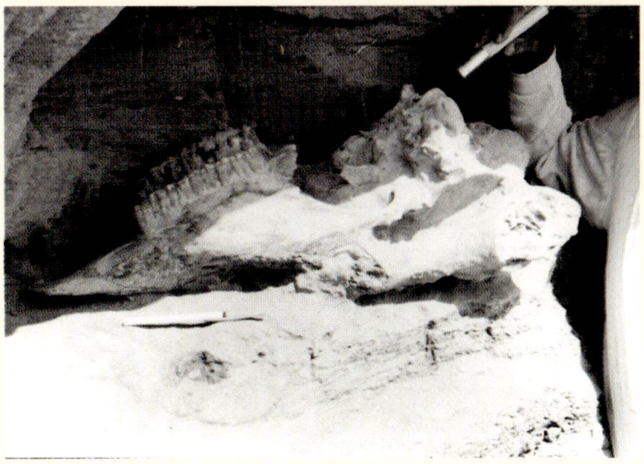

Pl. 3.10: A complete skull of *Bos namadicus* in the section at pilot trench. Note the position of the skull.

Pl. 3.11: The section where the skull of *Bos namadicus* was excavated in the pilot trench.

Pl. 3.12: An almost complete skull of *Bos namadicus* found *in situ* in the basal portion of the section at Guarighāṭ, downstream of the pilot trench.

Pl. 3.13: A partial skull of *Bos namadicus* in channel gravel deposit.

Pl. 3.14: An *in situ* complete rib of a mammal in the section near Balaghāṭ.

Pl. 3.15: A skull of *Bos namadicus* excavated from a silty horizon at Luknu on way to Umeria, north of Devakachar. The horncores are complete. However, the front portion of the skull, which was buried in the soil, was damaged.

Pl. 3.16: A complete right horncore of *Bos namadicus*. The left horncore is buried in the gravelly horizon.

Pl. 3.17: A horncore from the top of clayey horizon in one of the gullies northwest of the pilot trench near Gurāriāghāṭ.

Pl. 3.18: A view of Sūkhā Nālā near Devakachar where a mandibular symphysis of *Elephas* was excavated from the base of the section. A non *in situ* ash deposit was also found in the gravel in the top portion of the section.

Pl. 3.19: Close up of the mandibular symphysis of *Elephas* (the symphyseal region got damaged during excavation).

Pl. 3.20: A trial trench south of the pilot trench towards the Sher-Umer; ten such trial trenches were dug to ascertain the lithological characteristics and thickness.

Pl. 3.21: An excavated trench which yielded horncores of *Bubalus* sp.

Pl. 3.22: Horncores of *Bubalus* sp. from 3.21 excavated trench

Pl. 3.23: A section on the left bank of Narmadā near Barmanghāṭ below the PWD Rest House where an almost complete burial of a huge buffalo was excavated.

Pl. 3.24: A lower portion of the dolerite dyke at Barmanghāṭ where a large number of Stone Age tools were collected in addition to some fossils, preserved as lag deposits after the fluvial transport.

Pl. 3.25: A thick section of the alluvium near Jhānsighāṭ (on way to Jabalpur) where a lot of faunal and cultural material was found.

Pl. 3.26: A general view of the Narmadā at Hathnorā (northeast of Hoshangabad). The bouldery gravel bed (sheet deposit) yielded a partial cranium of *Homo erectus narmadiensis* in 1984. It is a GSI find.

Pl. 3.27: Section at Hathnorā. The pointer on the right lower end is exactly the spot where the cranium was discovered in the basal gravelly bed.

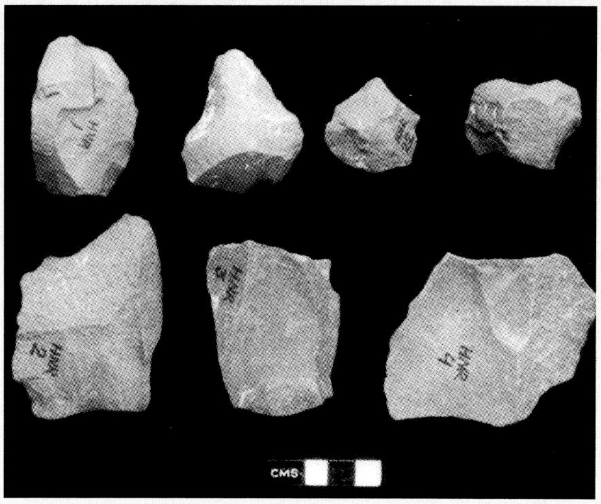

Pl. 3.28: Stone Age tools discovered at Hathnorā.

Pl. 3.29: Stone Age tools discoversed at Hathnorā.

Pl. 3.30: Stone Age tools discovered during the excavations near Jabalpur.

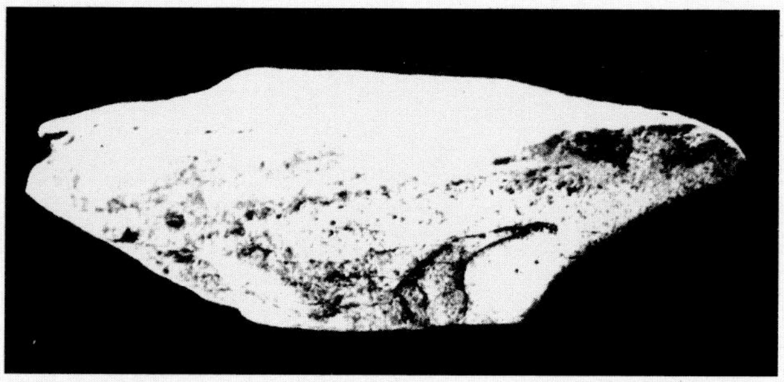

Pl. 3.31: Worked fossil bone from Gurāriāghāṭ.

Pl. 3.32: Stone Age tools collected from excavations in Narsinghpur District.

Pl. 3.33: Stone Age tools collected from excavations in Narsinghpur District.

Pl. 3.34: Kombewa flakes from Gurāriāghāṭ excavations.

The Kombewa method has mainly been reported from Africa. "A method of obtaining a flake of a form predetermined by the convexity of the lower face of another flake, previously knapped to serve as a core. An unretouched Kombewa flake thus possesses two lower surfaces. " Usually the purpose was to produce cleavers from the big Kombewa flakes of the African Acheulian (Inizan, M.L., H. Roche and J. Tixier 1992. *Technology of Stone.* Meudon: Crep).

Note: The above information was supplied by Dr Bishnupriya Basak and the author is grateful to her for this help.

Pl. 3.35: Some of the team members.

Pl. 3.36: Relaxing after a day's hard work.

Pl. 3.37: Fossil repository, cleaning and cataloguing in progress.

the author also completed a Ph.D. thesis on the archaeology and palaeontology of the present and neighbouring areas around Devakachār (Salahuddin, 1989) and another student studied the taphonomy of a part of fossils from the Narsinghpur Distt. (Kumar, 1999).

The present project gave us a better opportunity to excavate and explore several areas around Devakachār. Though there are little variations in stratigraphical units and their thickness on the right bank of the streams, there is predominance of silty and clayey units in these tributaries as stated earlier. The excavation conducted at Gurāriāghāṭ (opposite to the village Devakachār) and carried out for three seasons, was considered as the pilot trench. It exposed a complete, undisturbed sequence of lithounits dug in the form of step trenches. A number of trenches were excavated on other sides of the section, laterally, on the back side, in the gulleys and towards the Sher-Umer, to ascertain the continuity or otherwise of lithostratigraphy of the area.

The main section of the type area is about 30 m thick starting with a 5 m thick red brown concretionary silt at the base. This is followed by sandy pebbly gravel of 8 m thickness and thence by yellow brown concretionary silt about 18 m thick being covered finally by Black Cotton Soil. This section along with others in the general area is most prolific in faunal and cultural materials.

Another prolific section (Guarighāṭ) is located about 50 m downstream of the pilot trench at Gurāriāghāṭ. This section exposed in the form of an amphitheatre (because of uneven erosion of the alluvium in the central portion) has yielded a number of fossils and Stone Age tools. Here the section of about 25 m exposes from the base a 5 m thick red brown concretionary silt followed by a 5 m thick sandy pebbly gravel, the lower part of which is cross-bedded. A unit of yellow brown concretionary silt of 15 m thickness covers the alluvium which is in turn capped by Black Cotton Soil.

At Umeriā (about 1 km north of Devakachār) the fossiliferous section has a thickness of 6 m being distinguishable into four lithounits, starting from the base with reddish clay of 2.70 m thickness, cross-bedded sandy pebbly horizon of 2 m thickness, yellow sandy silt 0.80 m thick and Black Cotton Soil 0.50 m thick. Though no cultural material has been collected here, the section yielded numerous fossils.

A trench was dug in Luknu section near one of the dried up *nālās* north of the pilot trench on way to Umeriā. Discovery of a fragment of horncore prompted us to dig a 3 m × 3 m trench in a section of 9 m thickness in the clay bed. Several days' digging at the site yielded an almost complete skull, with both horncores intact, of a huge buffalo which had probably slipped in a pool and later got fossilized. The front portion of the skull was, however, damaged.

A section was dug near the top of a silty clayey exposure in the Gurāriā Nālā north-west of the pilot trench and this yielded a complete skull (though ill preserved) with horncores of a fossil ox.

The stratigraphy of excavated sections in and around the pilot trench consisted of, from the base, mottled clay (about 5 m from the level of the river), followed by friable sandy beds with dark bands in between. This merged with reddish sandy horizon, the lower unit of which was cross-bedded followed by mollusc-bearing gravel, fine grained gravelly horizon and the Black Cotton Soil. Several mud pebbly bands were also present in the section at varying heights. In fact, this sequence also indicated evidence of flood deposits. Soil samples were collected from various lithounits in the section and chemically analysed, which threw important light on the environment of deposition of the Narmadā when the sediments were being laid down.

BARMANGHĀṬ (*fig.* 3.7; *Pls.* 3.23 & 3.24)

This locality, situated about 25 km to the north-west of Narsinghpur, is one of the richest implementiferous sites in the Narmadā Valley. A well preserved section is exposed below the Forest Rest House which comprises from the base, well cemented, cross-bedded sandy pebbly gravel about 5 m thick, being capped by 23 m thick deposit of horizontal clay and yellow brown kankary silt and has yielded several fossils and Middle Palaeolithic tools. Much of the cultural material is, however, concentrated in the sandy pebbly gravel. Incidentally, a few years ago fossils of what turned out to be the largest buffalo from South Asia were dug out by the author from the yellow brown silt in the same section with concentration of its parts being between 10 m and 13 m. It was almost a complete burial, being represented by skull, jaws, several vertebrae, limb fragments, rib fragments, carpal and tarsal bones.

A few metres downstream from this section is a Lower Palaeolithic site located by the side of a huge outcrop of a dolerite dyke. A large number of tools as also a few fossils (obviously transported) were collected from this site, most of the former made on dolerite.

JHANSIGHĀṬ (*Pl.* 3.25)

This area displays one of the most impressive sections on the left bank of the main channel of the Narmadā, about 28 km west of Bheḍaghāṭ. A

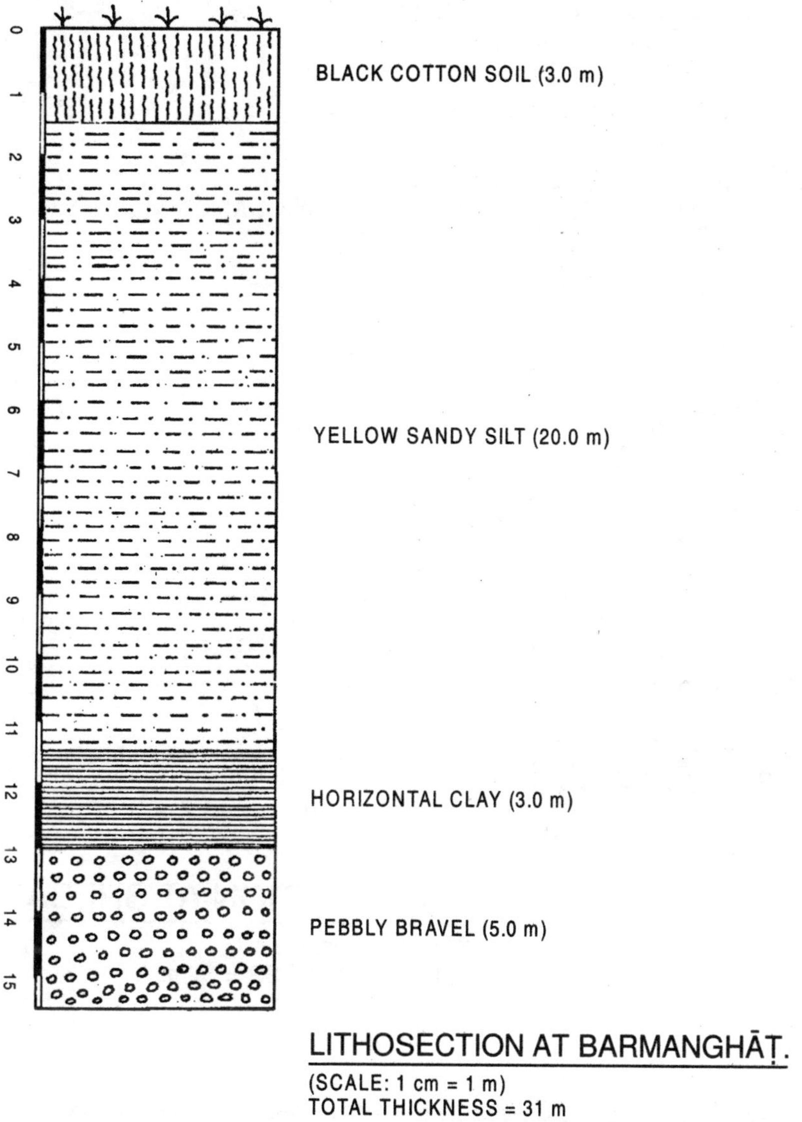

fig. 3.7: Lithosection at Barmanghāṭ

thick alluvium comprising red brown concretionary silt of 30 m or more is exposed here which has yielded Middle Palaeolithic tools and a few fossils, the former from the gravels and the latter from both the gravels and the overlying silts.

BHEḌAGHĀṬ (*fig.* 3.8)

Situated about 13 km south-east of Jabalpur this place is famous for a 9 m waterfall within the beautiful exposures of marble rocks. It is here that the Narmadā starts to flow in the form of a saucer shaped trough. The exposed section on the now dry and abandoned channel (or palaeo-channel) of the Narmadā above the waterfall, is about 52 m thick. The basalmost 2 m thick horizon consists of red clayey, non-kankary fissured

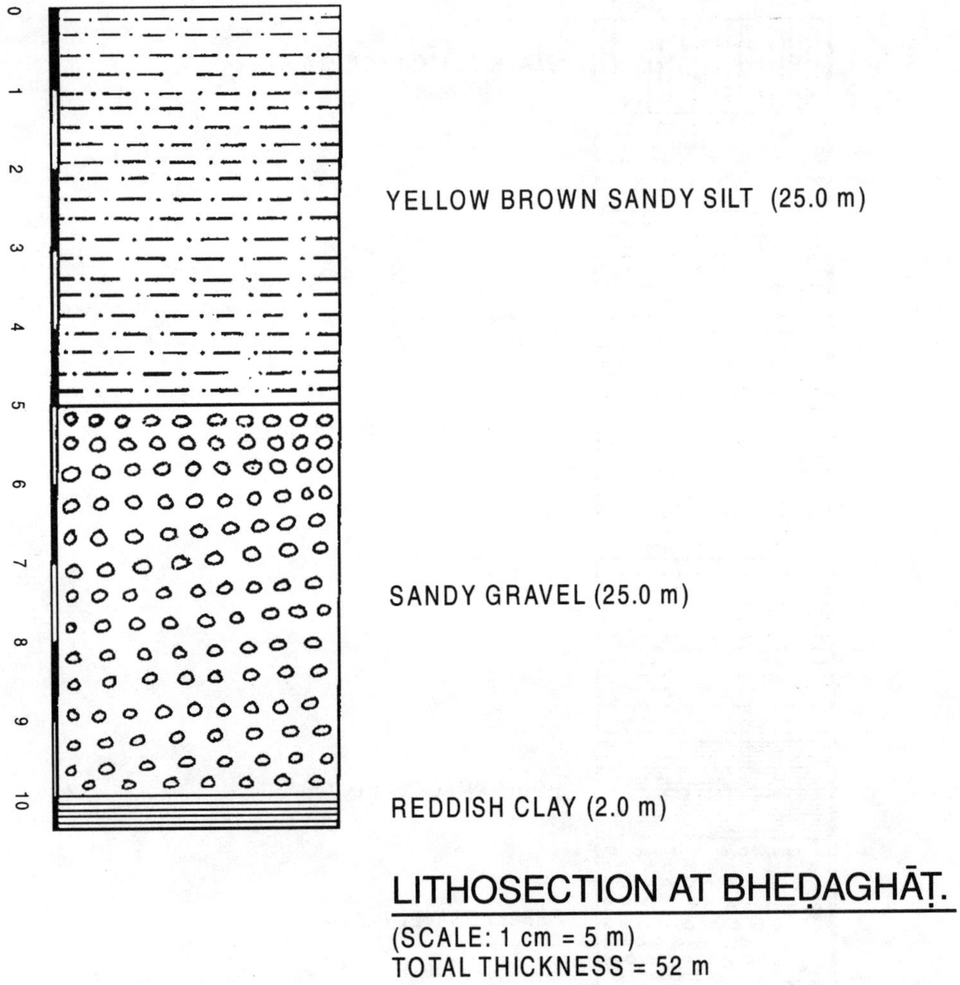

fig. 3.8: Lithosection at Bheḍaghāṭ

silt which is uniformly exposed in the entire region. This is disconformably overlain by sandy pebbly gravel about 25 m thick, the lower half of which is at places cross-bedded. This gravel is rich in chert and chalcedony pebbles and is overlain by a yellow brown concretionary silt of about 25 m thickness. This is one of the thickest sections in the Narmadā Valley having yielded Middle Palaeolithic tools and several fossils.

Composite Stratigraphy

Record of a complete stratigraphy is preserved at various sections (as at Hathnorā). In general Boulder Conglomerate forms the basal unit of the exposed sections. The pebbles and cobbles are fairly rounded and are of varying dimensions. They are well cemented, non-laminated and intercalated with fine sand and silt. At places where the Narmadā river flows in close proximity to Vindhya and Bijawar hills, huge concentration of gravel deposits is observed. The gravels have yielded both Lower and Middle Palaeolithic tools (made on sandstone and chert) consisting of choppers, scrapers, handaxes, cleavers, flakes and cores. The red brown concretionary silt, which conformably overlies the gravel bed, has variable thickness and a large lateral extent.

In general, a large number of fossils have been collected from the junction of the red brown concretionary silt and the overlying sandy pebbly gravel. There is an unconformity between these two horizons. However, the fossils are better preserved here than the ones collected from the underlying Boulder Conglomerate horizon. Perhaps the top surface of the reddish silt represents a stabilized landscape and this surface with a galleria forest cover provided an ideal habitat for wild animals. Some of these possibly got trapped in torrential monsoon floods of the subsequent period and a few got fossilized.

The thickness of the sandy pebbly gravel varies from section to section, its maximum thickness being 13 m at Bheḍaghāṭ. This unit is generally better sorted than the Boulder Conglomerate and is rich in chert and chalcedony minerals. Distinct cross-bedding is seen in the sandy portions of the gravel unit. The physical characters of the gravel and its inter-relation with underlying red brown concretionary silt suggest its deposition as a channel-bar over the ancient land surface. As said earlier, a large number of well preserved fossils have been collected from this unit in addition to Middle Palaeolithic tools.

The sandy pebbly gravel unit passes into yellow brown concretionary silt which is more sandy and less kankary than the red brown concretionary silt. This silt, at places, shows faint banding and cross-bedding. The presence of fine sandy lenses, clayey patches and fine laminations indicate that deposition occurred in a flood plain environment. The top surface of the silt has weathered into Black Cotton Soil.

In comparison with other peninsular rivers (Godāvarī, Kṛṣṇā, Bhīmā, etc.) the Narmadā Valley has preserved better and longer stratigraphical record of the Late Middle Pleistocene to Upper Pleistocene time bracket.

The lithostratigraphy of the Central Narmadā Valley as suggested by various authors is tabulated in Table 1. The chrono- and cultural stratigraphy of the Central Narmadā Valley as based on the new data is given in Table 2. Some of the numerous Stone Age tools collected during the present research are shown in *Pls.* 3.28 to 3.34 with the exception of *Pl.* 3.31 which is a worked fossil bone.

Table 1 : Lithostratigraphy of the Central Narmada Valley

Theobald (1860)		de Terra and Paterson (1939)			Khatri (1966)	Supekar (1968)	Badam (1979)	Biswas and Dassarma (1981)
Pliocene		Middle Pleistocene			Holocene to Middle Pleistocene	Early Upper Pleistocene	Early Upper to Late Middle Pleistocene	Upper Pleistocene
Upper Group	Black Cotton Soil Pale brownish alluvium Irregular beds of sands and conglomerates	Younger Alluvium	Cotton Soil Group	Black Cotton Soil Basal gravel --disconformity--	Yellow silt and modern soil Yellow cross-bedded sand Cemented sandy gravel	Black Cotton Soil Yellow brown concretionary silt Sandy pebbly gravel	Dark brown consolidated silt and gravel	Black Cotton Soil Grey silt Pedocol layer Grey sand with lime concretions
		Older Alluvium	Upper Group	Upper pink concretionary clay Upper gravel and sand --disconformity--	Red clay/ Boulder Conglomerate	Red brown concretionary silt Boulder Conglomerate	Yellow brown calcreted silt, sand and gravel Reddish brown calcareous silt Cemented boulder gravel	Pinkish clay and silt lime concretions Grit and sand Boulder bed
Lower Group	Red yellow clay Boulder Conglomerate		Lower Group	Lower red concretionary clay Basal conglomerate --disconformity-- Laterite gravel and soil	Laterite			Deep red clay Deep red boulder bed

Table 2 : Chrono- and Cultural Stratigraphy of the Central Narmadā Valley

Approximate age	Lithological units	Approx. thickness	Pal. and Arch. materials
Early-Mid. Holocene (15,000 to 4000 BP)	Dark brown moderately consolidated silt and gravel	-	Semi-mineralized bones, Mesolithic
Terminal Pleistocene (15,000 to 10,000 BP)	Reddish clay band	-	Upper Palaeolithic tools
Late Upper Pleistocene (40,000 to 15,000 BP)	Yellowish brown calcreted sands and gravels	10-13 m	Middle Palaeolithic tools and Upper Pleistocene fauna
Late Middle Pleistocene to early Upper Pleistocene (150,000 to 40,000 BP)	Reddish brown calcareous silt and well-cemented boulder gravels	3-8 m 15 m (at Hathnorā)	Lower Palaeolithic + Middle Palaeolithic tools + Mid. to late Pleistocene fauna *H.e. narmadiensis* and late Acheulian artefacts

Notes

1. A date ranging from 280,000—127,000 yrs. BP has recently been proposed for the Middle Pleistocene.
2. The hominid cranium is now dated to around 2,00,000 yrs. BP

4

Faunal Material

Previous Collections

NEXT only to the Śiwāliks of north-west India, the Narmadā Valley has yielded the richest mammalian fossils found in India. Some of them are well preserved and generally stained red or yellow, because of the presence of iron oxide minerals in the enclosing sediments, particularly at Varurewa, Sher and Umer streams in Narsinghpur Dist. Many of the fossils are found in association with Stone Age tools, making the Narmadā Valley very significant as far as the origins of man are concerned. Added to this is the important discovery of *Homo erectus narmadiensis* (*Pl.* 4.1 — *Pl.* 4.3) collected recently from Hathnorā (Sonakia, 1984).* The forests of the Narmadā Valley even today abound in large cats, bear, gaur, deer, black buck, reptiles, etc. Many of the ancestors of these animals lie buried in the Pleistocene deposits along with other mammals such as elephant, hippopotamus, horse, rhinoceros, cattle, deer etc.

The first authentic list of the vertebrate fossils from the valley was given by Pilgrim (1905), who treated the Narmadā and the Godāvarī valleys as of equivalent age. The fauna listed by Pilgrim is as follows:

Ursus namadicus, Bubalus palaeindicus, Leptobos frazeri, Bos namadicus, Cervus duvauceli, Sus sp., *Hippopotamus namadicus, Hippopotamus palaeindicus, Equus namadicus, Rhinoceros unicornis, Elephas namadicus, Stegodon insignis, Stegodon ganesa* and *Pangsura tectum.*

Most of the above mentioned genera except three (*Leptobos, Stegodon, Pangsura*) continue till the present day. *Rhinoceros unicornis* is identical with living species. *Bubalus palaeindicus* and *Cervus duvauceli* are closely related to the modern Indian buffalo and the ***bārāsiṅghā*** (the swamp deer) respectively. The turtles like *Pangsura tectum* are also similar to the turtles living in the Indian rivers today. The Indian cattle is a progenitor of *Bos namadicus.* Therefore, judging from the faunal assemblage, which is quite evolved, part of the Narmadā alluvium may be assigned to the later part of the Pleistocene.

---

* Additional hominid fossils have recently been discovered from the same locality.

de Terra and Paterson (1939) gave the following list of the fauna:

Elephas namadicus, Equus namadicus, Hexaprotodon namadicus, Bos namadicus, Bubalus palaeindicus, Sus sp., *Trionyx* sp., and *Emys* sp. They opine that the fauna of the Upper Narmadā Group is similar to that of the Lower Narmadā Group and that both carry Middle Pleistocene elements. But for a few fossils like *Leptobos* and *Stegodon,* the assemblage of fossils from Narmadā is typical of Middle Pleistocene of Eurasia (Hopwood, 1935) and younger than the Piñjor fauna of the Śiwāliks.

Colbert (1942) added two more species, *Holarctos namadicus* and *Palaeoloxodon namadicus* to the above faunal list and agreed with Pilgrim regarding the chronological position of the Narmadā deposits.

Hooijer (1963), on the basis of a preliminary identification of some faunal material from the Narmadā Valley gave the following list of fossils:

Elephas namadicus/Elephas maximus, Hexaprotodon namadicus/ Hippopotamus palaeindicus, Equus namadicus, Bos/Bubalus and *Sus* sp. According to him the collection is not sufficiently characteristic for an age determination; it may be Middle but more likely Upper Pleistocene.

Hooijer (1963) is of the opinion that the elephants in the collection would seem to be either an advanced *namadicus* or to belong to the modern species, *maximus.* The suid molar resembles Middle Pleistocene suids described from China, Burma and Java. The hippopotamus molar is either *Hexaprotodon namadicus* or the more advanced *Hippopotamus palaeindicus.* The long protocones in the teeth of horse collection agree with those of *Equus namadicus,* which is regarded as ancestral to the modern *Equus hemionus.* The bovine teeth may belong either to *Bos* or to *Bubalus.*

Khatri (1966) identified the following species of fossils from the Narmadā Valley:

Elephas namadicus (*antiquus*), *Elephas maximus* (*indicus*), *Hexaprotodon namadicus, Bos namadicus, Bubalus palaeindicus, Bison, Equus namadicus, Sus* sp. and *Cervus* sp. According to him the fossils range from Middle Pleistocene to Holocene and the degree of fossilization varies from horizon to horizon. He further suggested that some of the fossils from the Lower Group of Narmadā can be compared with those from the Boulder Conglomerate zone of the Śiwāliks.

Supekar (1968) added *Elephas hysudricus, Bubalus bubalis* and *Equus caballus* to the list of the Narmadā fauna. However, the present author has re-identified the equid skull in the collection as that of *Equus namadicus.*

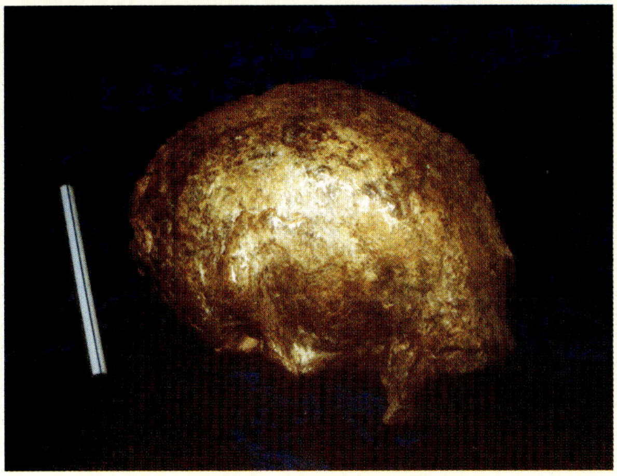

Pl. 4.1: Partial cranium of *Homo erectus narmadiensis* (photograph taken in GSI, Nagpur).

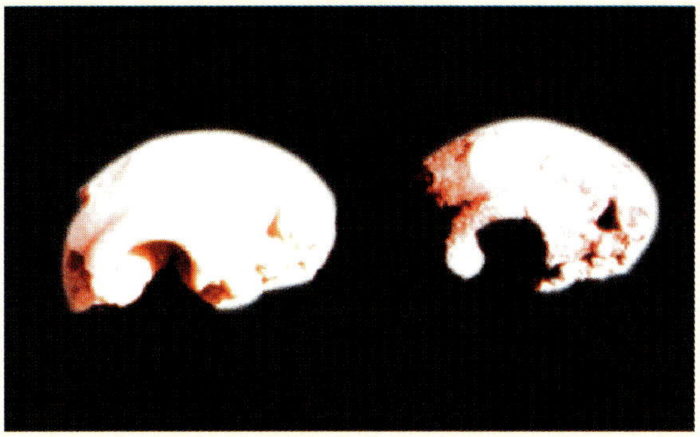

Pl. 4.2: Skulls of *Homo erectus narmadiensis* (fossil and cast).

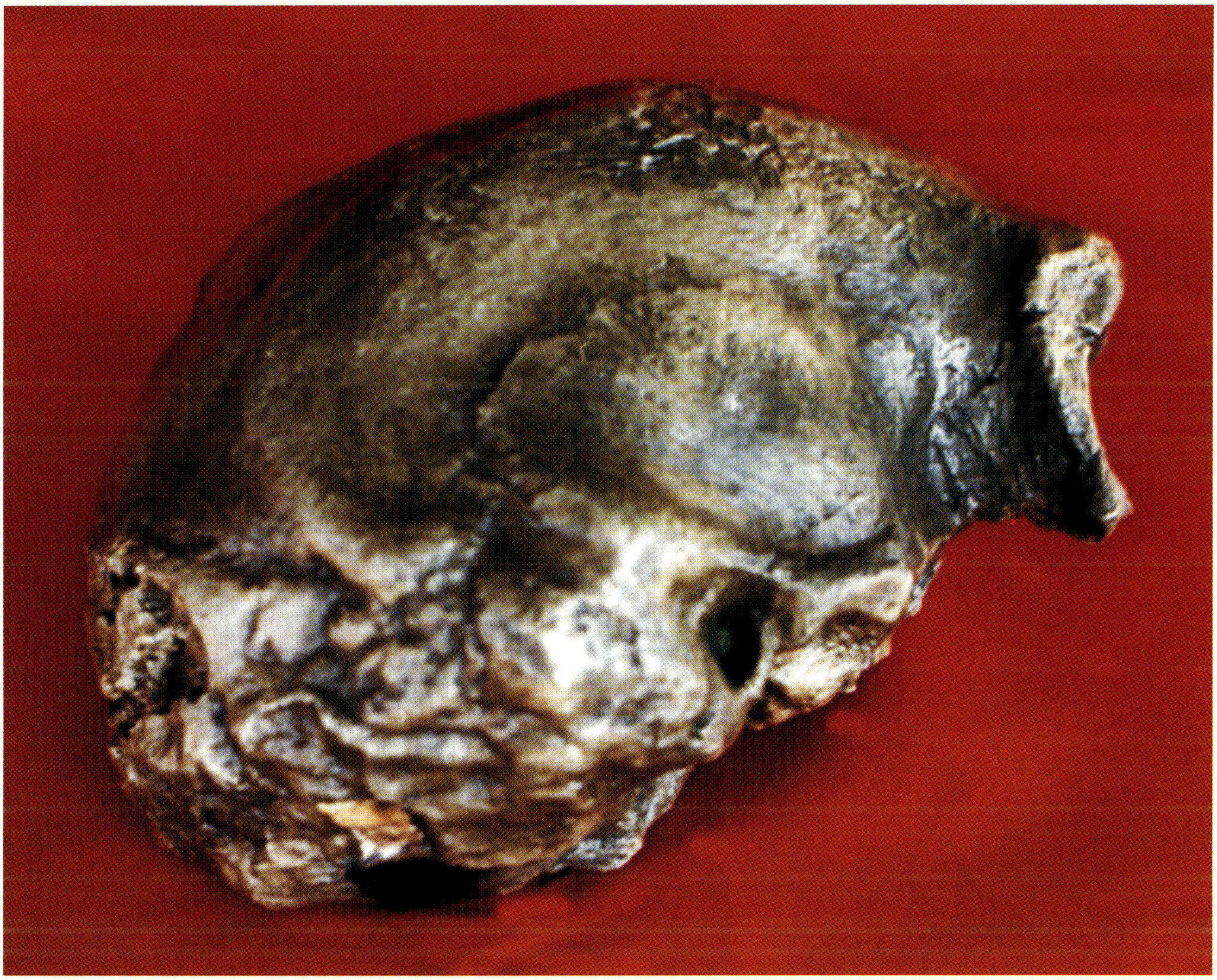

Pl. 4.3: Partial cranium of *Homo erectus narmadiensis* found at Hathnorā (cast).

Dassarma and Biswas (1978, 1979) have recorded a fossil mandible of a slightly advanced form *Hexaprotodon namadicus,* with six incisors but with second pair reduced in size, from the middle alluvial unit (Upper Pleistocene) and a fossil mandible of a much advanced form *Hippopotamus palaeindicus* with four incissors, similar to the present-day hippopotamus, but occasionally showing traces of the second pair of incisors of *Hexaprotodon.*

Badam (1979a, 1979b), in a detailed morphological analysis of the Narmadā fauna, confirms the presence of *Cervus duvauceli, Stegodon insignis-ganesa, Elephas hysudricus, Elephas namadicus, Hippopotamus palaeindicus,** *Hexaprotodon namadicus, Bos namadicus, Bubalus palaeindicus, Equus namadicus, Sus* sp., *Crocodylus palaeindicus, Gavialis* sp., *Trionyx* sp. The discovery of the reptiles, *Crocodylus, Gavialis* and *Trionyx* is of great significance in view of the scarcity of the reptilian fauna in the Narmadā Valley which is rather surprising.

The author is of the opinion, following Hooijer (1955), that *Stegodon ganesa* represents the male and *Stegodon insignis* the female of one and the same species. He agrees with the possibility that *Hippopotamus palaeindicus,* being more progressive than *Hexaprotodon namadicus,* may be confined to the Upper Pleistocene. The recent discovery of *Hexaprotodon palaeindicus* in the Ghoḍ Valley of Maharashtra (Badam and Kajale, 1977a, 1977b) dates it to about 20,000 BP which unquestionably places it within the Late Pleistocene.

Fauna Described from the Narmadā Valley by earlier workers is tabulated in Table 3.

Collection Under the IGRMS Project

As said earlier, a large number of fossils have been collected from the Central Narmadā Valley during the execution of the present project. Apart from megavertebrates and invertebrates (molluscan shells) (*Pl.* 5.58) which have been collected in course of earlier fieldwork also, microvertebrates have been discovered from the valley for the first time (*Pl.* 5.56) and hence this discovery represents the first report of microvertebrates from the Peninsular India with the exception of Kurnool Caves in Andhra Pradesh. Two papers on this discovery have already been published and an appendix on the findings is attached with this book. Fish vertebrae have also been collected (*Pl.* 5.57)

The megavertebrates were extensively collected. Though a complete list of these findings is given here, brief morphological descriptions of only

* This may be updated to *Hexaprotodon palaeindicus.*

Table 3 : Fauna from the Central Narmadā Valley in association with Lower and Middle Palaeolithic cultures

Order	de Terra and Paterson (1939)	Colbert (1942)	Hooijer (1963)	Khatri (1966)	Supekar (1968)	Badam (1979a, 1979b)	Badam and Salahuddin (1982)
Carnivora	—	*Holarctus namadicus*	—	—	—	—	—
Perissodactyla	—	*Rhinoceros unicornis*	—	—	—	—	—
	Equus namadicus	*Equus namadicus*	*Equus namadicus*	*Equus namadicus*	*Equus caballus*	*Equus namadicus*	*Equus namadicus*
Artiodactyla	*Hexaprotodon namadicus*	*Hippopotamus namadicus*	*Hippopotamus namadicus*	*Hippopotamus namadicus*	*Hippopotamus namadicus*	*Hippopotamus namadicus*	*Hippopotamus namadicus*
	—	*Hippopotamus palaeindicus*	*Hippopotamus palaeindicus*	—	—	*Hippopotamus palaeindicus*	*Hippopotamus palaeindicus*
	Bos namadicus	*Bos namadicus*	*Bos* sp.	*Bos namadicus*	*Bos namadicus*	*Bos namadicus*	*Bos namadicus*
	Bubalus palaeindicus	*Bubalus palaeindicus*	*Bubalus* sp.	*Bubalus palaeindicus*	*Bubalus palaeindicus* or *B. bubalis*	*Bubalus* sp. (probably *B. palaeindicus*)	*Bubalus* sp.
	—	*Leptobos frazeri*	—	—	—	—	—
	—	—	—	*Bison* sp.	—	—	—
	—	*Cervus duvauceli*	—	*Cervus* sp.	—	*Cervus duvauceli*	*Cervus duvauceli*
	Sus sp.	*Sus namadicus*	*Sus* sp.	*Sus* sp.	—	*Sus* sp.	—
Proboscidea	—	*Stegodon insignis*	—	—	*Stegodon insignis*	*Stegodon insignis ganesa*	*Stegodon insignis*
	—	*Stegodon ganesa*	—	—	—		—
	Elephas	*E. namadicus*	*Elephas namadicus* or *E. maximus* (*indicus*)	*Elephas namadicus* (*antiquus*), *E. maximus*	*Elephas* *E. hysudricus* *Elephas namadicus*	*Elephas maximus*, *Elephas hysudricus*	*Elephas* sp.
	—	*Palaeoloxodon namadicus*	—	—	—	—	—

Reptilia	—	—	—	—	—	*palaeindicus*	*Crocodylus*
	Trionyx sp.	—	—	—		*Trionyx* sp.	
	Emys sp.	—	—	—	—	*Emys* sp.	—
	—	—	—	—	—	*Gavialis* sp.	—

Notes

1. Collection made by the author subsequent to 1982 and prior to the commencement of the project: *Bibos gaurus*, *Bubalus* n. sp., *Antilope cervicapra*, *Axis axis*, *Boselaphus tragocamelus*, etc. are not included in the table.
2. The nomenclature of *Hippopotamus namadicus* and *Hippopotamus palaeindicus* is now updated to *Hexaprotodon namadicus* and *Hexaprotodon palaeindicus* respectively.

a few vertebrates are included here. A detailed morphological study will be taken up in due course for a specialised paper. Photographs and sketches of some of the fossils not included in morphological analysis have also been added.

A checklist of fossil vertebrates discovered under the project is given in Table 4.

Abbreviations

LOCALITY

BMGT : Barmanghāṭ

BLGT : Belghāṭ

BGT : Bheḍaghāṭ

DVK : Devakachār

DHNS : Dhansi

GGT : Guārīghāṭ

GUGT : Gurāriāghāṭ

HNR : Hathnorā

KDG : Kadgāon

KKG : Kiriāghāṭ

RTK : Ratikarar

TGT : Talayaghāṭ

UMR : Umeriā

Geological Unit:

SPG : Sand pebbly gravel

RBCS : Red brownish concretionary silt (including yellow brown concretionary silt)

Notes

1. In case of the fossils not collected from excavations, the attached matrix gave some clues to their provenance. However, some amount of uncertainty may exist in a few cases, e.g. Specimen Nos. NMD/1178/DC to NMD/1198/DC except NMD/1180/DC and NMD/1181/DC (see Table 4).
2. All the catalogued fossils are deposited with the IGRMS Repository.
3. Some of the sites are very close to each other, e.g. Umeriā and Devakachār and therefore sometimes these sites are treated as one.
4. Nomenclature of the fossils has generally been retained as given by the original authors except in a few cases.

Table 4 : Checklist of Fossil Vertebrates Discovered During the IGRMS Project

S.No.	Catalogue No.	Osteological/Dental Parts	Locality	Geological Unit	Genus/Species
1.	NMD/1067/DC	Skull	GUGT	RBCS	*Bos namadicus*
2.	NMD/1068/DC	Rib: complete	do	do	*Bos namadicus*
3.	NMD/1069/DC	Carapace fragment	do	do	*Trionyx* sp.
4.	NMD/1070DC	Humerus : left	do	do	Turtle
5.	NMD/1071/DC	Metatarsal shaft	DHNS	SPG	*Equus* sp.
6.	NMD/1072/DC	Maxilla : Rt. with M3	UMR	do	*Hexaprotodon* sp.
7.	NMD/1073/DC	Rib fragment	RTK	do	A large mammal
8.	MD/1074/DC	Calcaneum : left	do	do	*Bos namadicus*
9.	NMD/1075/DC	Scaphoid	TGT	do	*Hexaprotodon palaeindicus*
10.	NMD/1076/DC	Metatarsal, right	do	do	*Bos namadicus*
11.	NMD/1077/DC	Ramus : Lt. M2&M3	do	do	*Bos namadicus*
12.	NMD/1078/DC	Ramus : Rt. erupting M1 & M2	TGT	do	*Bos namadicus*
13.	NMD/1079/DC	Metatarsal	BGT	do	*Antilope cervicapra*
14.	NMD/1080/DC	Canine fragment	UMR	do	*Hexaprotodon* sp.
15.	NMD/1081/DC	Canine fragment	do	do	*Hexaprotodon* sp.
16.	NMD/1082/DC	Canine fragment	do	do	*Hexaprotodon* sp.
17.	NMD/1083/DC	Rib fragment	KDG	do	A large mammal
18.	NMD/1084/DC	Left upper M1/M2	TGT	do	*Bos/Bubalus*

19.	NMD/1085/DC	Tibia : proximal, left	do	do	*Bubalus palaeindicus*
20.	NMD/1086/DC	Acetabulum : left	do	do	*Hexaprotodon* sp.
21.	NMD/1087/DC	Humerus : distal, left	do	do	*Bubalus palaeindicus*
22.	NMD/1088/DC	Cervical : centrum	DVK	do	A large mammal
23.	NMD/1089/DC	Lamellar fragment	do	do	*Elephas* sp.
24.	NMD/1090/DC	Tibia shaft	UMR	do	A large mammal
25.	NMD/1091/DC	Humerus : proximal, right	RTK	do	*Hexaprotodon* sp.
26.	NMD/1092/DC	Maxilla : left with M2 and M3	UMR	do	*Hexaprotodon* sp.
27.	NMD/1093/DC	Lumbar vertebra	TGT	do	*Hexaprotodon palaeindicus*
28.	NMD/1094/DC	Axis vertebra	HNR	do	A large mammal
29.	NMD/1095/DC	Femur : distal, left	TGT	do	*Bos namadicus*
30.	NMD/1096/DC	Humerus : shaft & head	HNR	do	*Elephas* sp.
31.	NMD/1097/DC	Mandible: left with molar	DVK	do	*Elephas namadicus*
32.	NMD/1098/DC	Lumbar vertebra	DVK	do	A large mammal
33.	NMD/1099/DC	Lower molar : Rt	GGT	do	*Elephas namadicus*
34.	NMD/1100/DC	Humerus : distal, left	TGT	do	*Bubalus palaeindicus*
35.	NMD/1101/DC	Acetabulum fragment	DVK	do	*Elephas* sp.
36.	NMD/1102/DC	Mandibular symphysis with cheek teeth	UMR	do	*Hexaprotodon palaeindicus*
37.	NMD/1103/DC	Vertebral fragment	HNR	do	*Elephas* sp.
38.	NMD/1104/DC	Mandible : right with molar	DVK	do	*Elephas namadicus*
39.	NMD/1105/DC	Rib fragment	TGT	do	*Elephas* sp.
40.	NMD/1106/DC	Cranium : frontal & ridge	BGT	do	*Bos namadicus*
41.	NMD/1107/DC	Mandible : right with molar	HNR	do	*Elephas hysudricus*

42.	NMD/1108/DC	Tibia : proximal, right	UMR	do	*Bos/Bubalus*
43.	NMD/1109/DC	Carapace fragment	DVK	do	*Trionyx* sp.
44.	NMD/1110/DC	Isolated broken tooth	GGT	do	*Elephas* sp.
45.	NMD/1111/DC	Metacarpal	DVK	do	*Rhinoceros* sp.
46.	NMD/1031/DC	Atlas vertebra	TGT	do	*Bubalus palaeindicus*
47.	NMD/1113/DC	Thoracic vertebra	GUGT	do	A large mammal
48.	NMD/1114/DC	Thoracic vertebra	do	do	do
49.	NMD/1115/DC	Tibia shaft	HNR	do	do
50.	NMD/1116/DC	Thoracic vertebra	KDG	do	do
51.	NMD/1117/DC	Cervical vertebra	HNR	do	do
52.	NMD/1118/DC	Rib fragment : proximal end	GUGT	do	do
53.	NMD/1119/DC	Tusk fragment	HNR	do	*Elephas* sp.
54.	NMD/1120/DC	Scapula fragment	do	do	Large bovid
55.	NMD/1121/DC	Radius : distal, right	DVK	do	*Bos namadicus*
56.	NMD/1122/DC	Atlas fragment	TGT	do	*Elephas* sp.
57.	NMD/1123/DC	Tusk fragment	TGT	do	*Elephas* sp.
58.	NMD/1124/DC	Femur : shaft & distal end, Rt.	DVK	do	*Bubalus palaeindicus*
59.	NMD/1125/DC	Femur : distal	do	do	A large artiodactyle
60.	NMD/1126/DC	Femur shaft fragment	do	do	do
61.	NMD/1127/DC	Skull fragment	do	do	do
62.	NMD/1128/DC	Skull fragment	do	do	An artiodactyle
63.	NMD/1129/DC	Mandibular condyle : right	UMR	do	*Hexaprotodon* sp.
64.	NMD/1130/DC	Mandibular fragment	UMR	do	do

65.	NMD/1131/DC	2nd Phalanx	GUGT	do	*Panthera* cf. *P. pardus*
66.	NMD/1132/DC	Femur : distal	GGT	RBCS	*Bos namadicus*
67.	NMD/1133/DC	Femur : shaft fragment	UMR	SPG	*Bos/Bubalus*
68.	NMD/1134/DC	Horncore fragment	GUGT	do	*Bos namadicus*
69.	NMD/1135/DC	Radius : distal, left	HNR	do	*Bos/Bubalus*
70.	NMD/1136/DC	Metacarpal : proximal	do	do	Bovid
71.	NMD/1137/DC	Femur: distal condyle	do	do	*Bos/Bubalus*
72.	NMD/1138/DC	Radius : distal fragment	GGT	do	*Bos/Bubalus*
73.	NMD/1139/DC	Humerus : proximal, right	RTK	do	do
74.	NMD/1140/DC	Atlas fragment	GGT	do	do
75.	NMD/1141/DC	Metacarpal : distal	GUGT	do	*Bos namadicus*
76.	NMD/1142/DC	Humerus : distal, left	RTK	do	*Bos namadicus*
77.	NMD/1143/DC	Femur : shaft	DVK	do	A large mammal
78.	NMD/1144/DC	Scapula : right	KDG	do	*Bos namadicus*
79.	NMD/1145/DC	Tusk fragment	HNR	do	*Elephas* sp.
80.	NMD/1146/DC	Radius right	KDG	do	*Bos/Bubalus*
81.	NMD/1147/DC	1st phalanx	UMR	do	*Elephas* sp.
82.	NMD/1148/DC	Rib fragment	KDG	do	A large mammal
83.	NMD/1149/DC	Cervical vertebra	UMR	do	do
84.	NMD/1150/DC	Incisor fragment	BGT	do	Bovid
85.	NMD/1151/DC	Dental fragment	GGT	do	*Crocodylus palaeindicus*
86.	NMD/1152/DC	Metacarpal	HNR	do	A large artiodactyle
87.	NMD/1153/DC	Rib fragment	TGT	do	A large mammal
88.	NMD/1154/DC	Antler fragment	HNR	do	Cervid

89.	NMD/1155/DC	Metatarsal : proximal, left	DVK	SPG	*Antilope cervicapra*
90.	NMD/1156/DC	Mandible : anterior, left	do	do	Bovid
91.	NMD/1157/DC	Left upper M1/ M2	do	do	*Bos/Bubalus*
92.	NMD/1158/DC	Skull fragment	do	do	A large mammal
93.	NMD/1159/DC	Right upper M1	BGT	do	*Bos/Bubalus*
94.	NMD/1160/DC	Radius : proximal, right	GUST	do	do
95.	NMD/1161/DC	Metatarsal	DVK	do	*Equus* sp.
96.	NMD/1162/DC	Thoracic fragment	do	do	A large mammal
97.	NMD/1163/DC	Right lower M1	do	do	Bovid
98.	NMD/1164/DC	Tibia : proximal, right	do	do	*Bos/Bubalus*
99.	NMD/1165/DC	Scapula fragment	do	do	do
100.	NMD/1166/DC	2nd phalanx : proximal	do	do	*Bubalus palaeindicus*
101.	NMD/1167/DC	Radius : proximal	do	do	*Bubalus palaeindicus*
102.	NMD/1168/DC	Femur : distal, right	GUGT	do	*Canis aureus ?*
103.	NMD/1169/DC	Isolated molar	HNR	do	*Elephas hysudricus*
104.	NMD/1170/DC	Left lower M3	RTK	do	*Bos namadicus*
105.	NMD/1171/DC	Incisor fragment	DVK	do	*Hexaprotodon* sp.
106.	NMD/1172/DC	Left maxilla : P4-M3	DVK	do	*Bos namadicus*
107.	NMD/1173/DC	Vertebral fragment	do	do	A large mammal
108.	NMD/1174/DC	Femur: unfused	do	do	*Bos/Bubalus*
109.	NMD/1175/DC	Rib fragment	do	do	A large mammal
110.	NMD/1176/DC	Lower tooth fragment	do	do	*Bos/Bubalus*
111.	NMD/1177/DC	Deciduous M1/M2; upper	RTK	RBCS	*Bos/Bubalus*

112.	NMD/1178/DC	Left ramus with M1 and M2	-	do	*Bubalus palaeindicus*
113.	NMD/1179/DC	Upper molar	-	do	*Elephas hysudricus*
114.	NMD/1180/DC	Right ramus with M3	UMR	do	*Hexaprotodon palaeindicus*
115.	NMD/1181/DC	Left mandible with P4, M1, M2	do	do	*Bos namadicus*
116.	NMD/1182/DC	Left ramus with M3	-	-	*Bubalus palaeindicus*
117.	NMD/1183/DC	Ramus Rt. with M2, M3	-	-	*Bos/Bubalus*
118.	NMD/1184/DC	Left ramus with M2	-	-	*Bos/Bubalus*
119.	NMD/1185/DC	Left mandible with P4-M3	-	-	*Bos namadicus*
120.	NMD/1186/DC	Right ramus with P4 and M1	-	-	*Bos/Bubalus*
121.	NMD/1187/DC	Right M3, lower	-	-	*Bos namadicus*
122.	NMD/1188/DC	Left lower molar	-	RBCS	*Elephas hysudricus*
123.	NMD/1191/DC	Cervical vertebra	-	do	*Bos/Bubalus*
124.	NMD/1193/DC	Femur : proximal, left	-	-	*Bubalus palaeindicus*
125.	NMD/1194/DC	Radius : distal, left	-	-	*Bos namadicus*
126.	NMD/1195/DC	Tibia : distal, right	-	-	*Bubalus palaeindicus*
127.	NMD/1197/DC	Humerus, right	-	SPG	*Ovis/Capra*
128.	NMD/1198/DC	Metatarsal III, right	-	RBCS	*Panthera* cf. *P.leo*
129.	NMD/1240/DC	Lower jaw with P3-M2	DVK	SPG	*Bos namadicus*
130.	NMD/1277/DC	Upper Jaw with P2-M2	GGT	SPG	*Equus namadicus*
131.	NMD/1239/DC	Part of maxilla	TGT	RBCS	*Crocodylus palaeindicus*
132.	NMD/1322/DC	Left lower molar	GUGT	RBCS	*Equus namadicus*
133.	NMD/1243/DC	Lower jaw	DVK	SPG	*Bos namadicus*
134.	Unnumbered	Skull	KKG	RBCS	*Hexaprotodon palaeindicus*

Specimens Included

Chapter 5	Chapter 6
NMD/1198/DC	NMD/1053/DC
NMD/1102/DC	NMD/1239/DC
NMD/1180/DC	NMD/13/DC
NMD/1072/DC	NMD/11/DC
NMD/1092/DC	NMD/1051/DC
NMD/1181/DC	NMD/713/DC
NMD/1077/DC	NMD/548/DC
NMD/1070/DC	NMD/597/DC
NMD/1187/DC	NMD/775/DC
NMD/1157/DC	NMD/808/DC
NMD/1095/DC	NMD/1322/DC
NMD/1074/DC	NMD/498/DC
NMD/1076/DC	NMD/1091/DC
NMD/1078/DC	NMD/1188/DC
NMD/1144/DC	NMD/1079/DC
NMD/1194/DC	
NMD/1121/DC	
NMD/1178/DC	
NMD/1182/DC	
NMD/1187/DC	
NMD/1193/DC	
NMD/1185/DC	
NMD/1078/DC	
NMD/1183/DC	
NMD/1186/DC	
NMD/1184/DC	
NMD/1177/DC	
NMD/1191/DC	
NMD/1179/DC	
NMD/1197/DC	
NMD/1179/DC	
NMD/1188/DC	
NMD/1099/DC	

Note: The above numbered fossils are either represented by photographs or sketches or by both. Refer to Table 4.

5

Systematic Palaeontology

A TOTAL of seventeen species of fossil vertebrates were collected from the Central Narmadā Valley under the IGRMS Project. These are represented by complete to poorly preserved crania, dental and other skeletal elements which include vertebrae, ribs, pectoral and pelvic girdles, long bones and their extremities, turtle carapaces, etc. This portion deals with the morphology of skeletal elements of some of the animals identified as under:

- *Panthera* cf. *Panthera leo* (lion)
- *Panthera* cf. *P. pardus* (leopard)
- *Canis aureus?* (wolf)
- *Elephas hysudricus* (extinct elephant)
- *Elephas namadicus* (extinct elephant)
- *Rhinoceros* sp. (rhinoceros)
- *Equus namadicus* (extinct horse)
- *Sus namadicus* (extinct pig)
- *Hexaprotodon palaeindicus* (extinct hippo)
- *Hexaprotodon namadicus* (extinct hippo)
- *Bos namadicus* (extinct cattle)
- *Bubalus palaeindicus* (extinct buffalo)
- *Antilope cervicapra* (black buck)
- *Ovis/Capra* (sheep/goat)
- *Cervus* spp. (deer)
- *Crocodylus palaeindicus* (crocodile)
- *Trionyx* sp. (turtle)

Due to the absence of diagnostic features some of the fossils could not be identified precisely. Palaeobiogeographical distribution and evolutionary history of a few taxa have been included here which throw some light on their early history in India with special reference to evolution

---

Notes: 1. Some of these taxa have been updated in this list.

2. Only a few of the specimens listed in Table 4 have been included for morphological analysis.

3. If more than one species of deer family are established, then the collection would make eighteen species.

and migration during the Pleistocene period. Detailed and traditional mophological descriptions have been avoided.

Panthera cf. *P. leo* Linnaeus, 1758

SYSTEMATIC PALAEONTOLOGY

Class	:	Mammalia Linnaeus, 1758
Order	:	Carnivora Bowdich, 1821
Family	:	Felidae Gray, 1821
Geuns	:	*Panthera* Linnaeus, 1758
Species	:	cf. *P. leo* Linnaeus, 1758
Locality	:	Gurāriāghāṭ
Geological unit	:	Red Brown concretionary silt

NMD/1198/DC: Right metatarsal III

The specimen is a complete third metatarsal with slightly abraided edges. Proximal end displays a mediolaterally directed articular surface. Shaft is flattened mediolaterally. Distal trochlea has a single pully like shape which is very pronounced.

This is the first report of this species from the Central Narmadā Valley. This species had earlier been reported from the Kurnool caves in Andhra Pradesh and the Gandheshwary river in West Bengal. However, these felines are poorly recorded in India during the Pleistocene period.

Hexaprotodon palaeindicus Falconer and Cautley, 1836

SYSTEMATIC PALAEONTOLOGY

Class	:	Mammalia Linnaeus, 1758
Order	:	Artiodactyla Owen, 1848
Family	:	Hippopotamidae Gray, 1821
Genus	:	*Hexaprotodon* Falconer and Cautley, 1836
Species	:	*Hexaprotodon palaeindicus* Falconer and Cautley, 1836

DIAGNOSIS

Small braincase, sagittal crest well developed, orbits high, skull short and wide; mandibular symphysis short in proportion to its width and higher in proportion to its length; jaws squarish, each jaw comprising three subequal incisors; 2nd incisors situated at higher elevation in relation to 1st and 3rd incisors (Hooijer, 1950; Badam and Kajale, 1977b).

TYPE

A lower jaw, Falconer and Cautley (1847), No. 19743.

LOCALITY AND GEOLOGICAL UNIT

Narmadā Alluvium, Pleistocene.

ADDITIONAL MATERIAL

NMD/1102/DC: Mandibular symphysis : *Hexaprotodon palaeindicus*

NMD/1072/DC: Right maxilla with M3 : *Hexaprotodon palaeindicus*

NMD/1092/DC: Left maxilla with M2 and M3 : *Hexaprotodon palaeindicus*

NMD/1180/DC: Right ramus with part of M3: *Hexapotodon palaeindicus*

UNDERSCRIBED MATERIAL

NMD/1082/DC: Canine fragment : *Hexaprotodon* sp.

NMD/1086/DC: Acetabulum, left : *Hexaprotodon palaeindicus*

NMD/1093/DC: Lumbar vertebra : *Hexaprotodon palaeindicus*

NMD/1129/DC: Mandibular condyle fragment : Rt : *Hexaprotodon* sp.

NMD/1171/DC: Incisor fragment : *Hexaprotodon* sp.

Un-numbered: Part of skull : *Hexaprotodon palaeindicus*

NMD/1102/DC : Mandibular symphysis

Locality : Umeriā

Geological Unit : Sandy pebbly gravel

The specimen is an almost complete mandibular symphysis. However, several freshly broken mandibular elements of the same specimen were found together at one place. Probably the symphysis was broken on account of sediment load. Canines and incisors are broken, exposing respectively their elliptical and circular shaped alveoli. There are a total of six incisors, three on each side of the symphysis and one canine each at the lateral ends of the mandibular symphysis. Symphysis is broad, long and rectangular in shape as seen anteriorly. It has a shallow depression below the point of division between the sets of incisors. The length of symphysis is more than its width. Almost all the premolars have lost their crown.

Molars are moderately worn. The height of mandible is higher than those of the other specimens in the collection which is one of the important characteristics of *H. palaeindicus*. Right first molar is severely damaged and only the anterior cusp is partially preserved. Left first molar is worn lingually while hypoconids and protoconids are oval-shaped and fairly preserved. Right second molar is well preserved and shows less wear. Enamel is thick, posterior cusp is partially covered with matrix. Protoconid is U-shaped and has a thick enamel. Metaconid is mesio-distally elongated and enamel is moderately crenulated. Cingulum can be seen distally. Hypoconid is partially covered with matrix. Entoconid and hypoconid are circular and show slight wear.

Left second molar is well preserved with a moderate wear. Metaconid is more elongated than protoconid and covers the mesiodistal area of the cusp. Protoconid is small, oval-shaped and oblique, bucco-lingually oriented. Dentine is exposed in the cusps. Hypoconid and entoconid are similar in shape and size to that of the right counterpart.

Both the third molars are complete and tooth wear is almost insignificant. Metaconid and protoconid are oval-shaped and located opposite to each other. Cingulum is weakly developed. Hypoconulids are located at the posterior-most ends of the jaws and represent a small portuberance next to the cusps of M3. On the whole, mandibles are deeply curved and exhibit marked concavity on their buccal sides. The right mandibular condyle is broken.

NMD /1180/DC: Right ramus with M3

Locality : Umeriā

Geological Unit : Sandy pebbly gravel

The specimen belongs to posterior-most end of the jaw with broken M3 and partly broken coronoid process. Broken edges of the ramus are quite fresh.

Molar crown is not completely developed. Proto- and metaconid are moderately developed and are respectively oval and circular shaped. Talonid is situated at the distal end of the jaw. Broken edge of the coronoid process is heavily abraided. There are two puncture marks below the mandibular hinge which call for detailed taphonomical study.

NMD/1072/DC: Right maxilla with M3

Locality : Umeriā

Geological Unit : Sandy pebbly gravel

The specimen represents only a small fragment of right maxilla with an intact 3rd molar. Tooth enamel shows considerable wear. Paracone and metacone are oval-shaped and have deep enamel folds giving an appearance of a four-petalled flower. Protocone is V-shaped whereas the paracone is W-shaped. Two cusplets can be seen in the middle of both anterior and posterior cusps of the tooth. The labio-lingual enamel folds to the extent that a sharp crescent is formed at the anterior-most aspect of the tooth.

NMD/1092/DC: Right maxilla with M2 and M3

Locality	:	Umeriā
Geological Unit	:	Sandy pebbly gravel

The specimen is represented by a part of maxilla and palate with completely worn 2nd molar and an intact 3rd molar the enamel of which is also partly worn out. Paracone is elongated and W-shaped while the protocone has a somewhat V-shaped enamel outline. Adjacent to these cusps, enamel folds run labio-lingually. Meta- and hypocone are oval-shaped and enamel of the former is worn out. Cingulum is mildly developed on the outer side of the hypocone. Broken edges of the maxilla are fresh.

REMARKS

The appearence of *Hexaprotodon* in India is dated to the Dhok Pathan Stage of the Middle Śiwāliks (Middle Miocene). In subsequent geological periods it has been found widely distributed and represented by several species some of them notably, *Hexaprotodon sivalensis* (Piñjor Śiwāliks of NW India), *H. namadicus* (Middle Pleistocene) and *H. palaeindicus* (Late Pleistocene) have been recognized in the Indian context, the last two in Central and Peninsular river valleys.

The genus *Hexaprotodon* is characterized by each jaw having three pairs of incisors; mandibular symphysis long; orbits less prominent; lacrymal bone in close contact with maxillary bone, being excluded from orbital rim by frontal; astragalus elongated (Falconer, 1859: 237; Lydekker, 1882: 28-33; Matthew, 1929:556). The bi-rooted condition of the first upper premolar is an important character in *Hexaprotodon* besides the presence of three pairs of incisors. However, this appears to have been a Miocene feature which has remained consistent only in the Asian *Hexaprotodon* of the Pleistocene.

The family Hippopotamidae is represented by several extinct species which have lived in Europe, Africa and Asia. However, there are only two living species, viz. *Hippopotamus amphibius* and *Hippopotamus liberensis* which are confined to sub-Sahara and Central Africa. The Śiwālik variety *H. sivalensis* is supposed to have been derived from African *Hexaprotodon*. *Hexaprotodon namadicus* has been considered an index fossil for the Middle Pleistocene (Badam, 1984) and is believed to have been derived from *H. sivalensis* (Hooijer, 1950).

Hexaprotodon palaeindicus, in recent years, has been reported from several river valleys in peninsular India during the later part of Upper

Pleistocene. On morphological grounds. *H. namadicus* and *H. palaeindicus* have been treated as independent species of which the latter is considered to have descended from the fomer (Badam and Kajale, 1977b; Biswas and Dassarma, 1984). However, in the light of minor differences in 2nd incisors, canines, cheek teeth measurements and indices of the mandibular symphysis, Salahuddin (1989) has advocated that both the species may be treated as synonyms to each other.

The present collection includes several maxillae, lower complete dentition, isolated teeth, vertebrae, acetabulum, long bones and metapodials of which most of the elements are well preserved. Evidences of abrasion and/or weathering in the collection are minimum. In a few cases, however, breakages are present and appear to be of post-fossilization phase. A complete skull of *H. palaeindicus* was collected during the present project research from the reddish silty clay in Kiriāghāṭ near Shahganj, a little away from Hoshangabad (*Pls.* 5.17 and 5.18).

Bos namadicus Falconer, 1859

SYSTEMATIC PALAEONTOLOGY

Class	:	Mammalia Linnaeus, 1758
Order	:	Artiodactyla Owen, 1848
Family	:	Bovidae Gray, 1821
Genus	:	*Bos* Linnaeus, 1758
Species	:	*Bos namadicus* Falconer, 1859

Bos namadicus, the most profusely found fossil in the Indian Pleistocene has cranium contracted between the horncores and orbits; cross-section of horncores circular or oval; premaxillae separated from the nasals; occipital crest forming a regular curved arch, its summit does not reach up as far as the middle of the horncores; supraoccipital crest exhibits considerable overhanging on the infraoccipital; temporal grooves deeply indent occipitals.

Its type is collected from "Narmadā Alluvium, late Mid Pleistocene."

Out of the collection listed below only a few are described here:

NMD/1067/DC	:	Skull: almost complete
NMD/1077/DC	:	Left ramus: M2 & M3
NMD/1078/DC	:	Right ramus: M1, M2 (erupting)
NMD/1181/DC	:	Left mandible: P4, M1 & M2
NMD/1187/DC	:	RM3: lower

NMD/1170/DC	:	LM3: lower
NMD/1177/DC	:	DM1/M2: upper
NMD/1157/DC	:	LM1/M2: upper
NMD/1095/DC	:	Femur : left distal
NMD/1074/DC	:	Calcaneum: left
NMD/1076/DC	:	Metatarsal: right
NMD/1144/DC	:	Scapula: right
NMD/1121/DC	:	Radius : right distal
NMD/1194/DC	:	Radius : left distal
NMD/1068/DC	:	Rib: complete
NMD/1106/DC	:	Cranial fragment
NMD/1134/DC	:	Horncore fragment
NMD/1141/DC	:	Metacarpal distal

NMD/1181/DC: Left mandible with P4, M1 & M2

The specimen is represented by an incomplete jaw broken at the medial end. Corpus mandible is also partially damaged. Fourth pre-molar is bucco-lingually flattended and has thick enamel folds which have not yet developed completely.

First molar is slightly broken at the anteriormost end of the cusp. It is rectangular in shape and basal pillar has attained the crown height. Lingual cusps are oval-shaped. Pre-and post-fossettes are medio-laterally extended. Second molar is larger and wider than the first. Buccal (V-shaped), and lingual (oval-shaped) cusps are very well defined. Basal pillar has not yet erupted up to the crown height. Median ribs are strong.

NMD/1077/DC: Left Ramus with M2 & M3

The specimen represents a well preserved left ramus with M2 & M3. Anterior cusp of second molar is exposed. There are a few cracks on the ramus which may be due to pre-burial weathering.

Lingual cusps of M2, viz. entoconid and metaconid are oval-shaped and the buccal cusps, proto- and hypocone are V-shaped. Pre- and post-fossettes are mesio-distally elongated and shallow. Basal pillar is damaged. Median ribs are strong. M3 is more elongated and has an accessory cusp which is situated at the anteriormost end of the mandible. Hypoconulid is bucco-lingually compressed and lacks a thick enamel fold which differentiates it from that of the buffalo. Basal pillar is poorly developed.

NMD/1170/DC: Lower M3

This is an isolated third molar with part of its abraided roots. Meta- and entoconids are predominantly oval-shaped. Pre- and post-fossettes are well developed, enamel is highly crenulated and anterior fossette has a small cusplet like enamel loop towards the posterior aspect of the cusp. Hypoconulid is small, oval-shaped and has a sharp enamel fold directed towards anterior end of the jaw. In contrast, this accessory cusp in buffalo is large, circular and with a sharp enamel fold.

NMD/1187/DC: Lower M3

The specimen is a well preserved third molar with part of ramus which has lost much of corpus mandible. Broken edges are abraided. Roots of the tooth are also exposed. Metaconid and entoconid are bucco-lingually elongated and oval-shaped, while the outer cusps (proto-and hypoconid) are U-shaped. Pre- and post fossettes are crenulated, moderately elongated and mesio-distally directed. Hypoconulid is small with a thinner enamel fold. Basal pillar is loop-shaped and exposed close to the crown height. The specimen is stained black indicating the presence of high proportion of manganese in the sediments enclosing it.

NMD/1157/DC: Upper LM1/LM2

This is an isolated tooth which has lost the roots, the broken edges are fairly abraided. Para- and metacones are oval-shaped while proto- and hypocones are V-shaped. Pre- and post-fossettes are widely spread over mesio-distally. Enamel of these fossettes is crenulated and somewhat thick. Basal pillar is strong, so are the median ribs.

NMD/1095/DC: Femur, distal left (*fig.* 5.1)

The specimen is represented by the distal end and part of shaft of femur. The state of preservation is fairly good. The broken edges of shaft are fresh. However, the epiphyseal ends of medial condyle exhibit some superficial pitting which may be attributed to the bacterial activity prior to burial. Supracondyloid fossa is deep and vertically elongated. It is situated above the lateral condyle. Patellar surface is narrow at the base and broadens anteriorly. Lateral and medial condyles are well represented. Intercondyloid fossa is broad and deepens below the patellar surface. However, the depth and width of the fossa is less than in *Bubalus*. Colour of the bone indicates that the sediments which yielded it were rich in iron oxide.

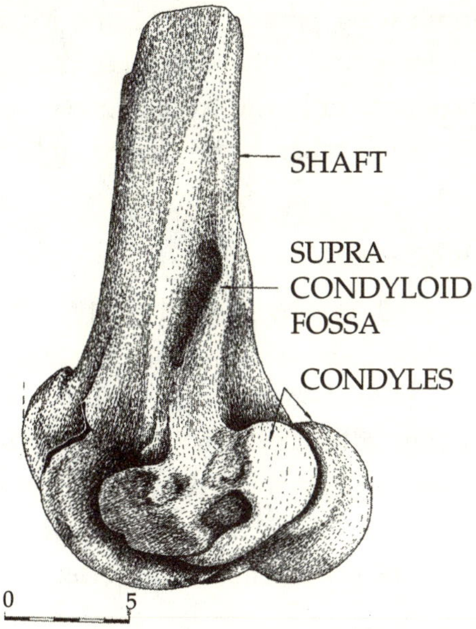

fig. 5.1: NMD/1095/DC: Distal half of femur of *Bos namadicus*

NMD/1074/DC: Calcaneum, left

The specimen is an almost complete calcaneum except for the anterior process. Corpus calcanei is elongated with a rounded head. However, the shaft of the corpus is laterally flattened. It terminates into sustentaculum tali and facies articularis talaris which is one of the diagnostic elements in the bone, helpful in differentiating *Bos* from *Bubalus*. Facies articularis is not far pushed backwardly. The facet of anterior process situated below is not sharply directed upwards, a character that helps to identify it as that of *Bos namadicus*.

NMD/1076/DC: Metatarsal, Right

The specimen is represented by an almost complete metatarsal except the distal end close to the nutritricium foramen. The proximal end of the bone represents a set of posterior and anterior articular grooves which receive naviculo-cuboid and first, second and third tarsals. The shape of these facets is an important diagnostic character in assigning it to the generic level. Facets for naviculo-cuboid are short, facet for lateral cuneiformis has a more irregular axial side. The facet for tarsal I shows irregular shape and is separated by a shallow groove. A deep broad median groove runs vertically throughout the bone.

NMD/1078/DC: Right ramus with M1 and M2

The ramus is poorly preserved with M1 and erupting M2. Broken edges of the ramus are slightly abraided. Surface of ramus shows some superficial mosaic cracks. The tooth wear pattern in M1 and M2 suggests a young age for the individual. Inner and outer cusps are not developed. Pre-and postfossettes are hollow and only the former preserves enamel outline. Median ribs are weak. Basal pillar is developed upto half of the crown height.

NMD/1144/DC: Scapula, Right

The specimen is represented only by glenoid cavity, collum and some part of the blade. Broken ends of the bone are fresh indicating a post-fossilization damage. Glenoid cavity is oval-shaped and shallow. Knob is located on the lateral aspect of the collum which is broad and dorso-ventrally flattened. Spine is broken. Infraspinous fossa is preserved partially. There are a couple of deep to superficial vertical cracks along the glenoid and collum which seem to be of pre-fossilization phase.

NMD/1194/DC: Radius, distal left

The specimen is a moderately abraided distal radius. Shaft is broken near the distal end. Articular facets are shallow. Styloid process appears like a protuberance which is also moderately abraided.

NMD/1121/DC: Radius, Right Distal

The specimen is represented by a well preserved distal end and some part of shaft showing spiral breakage. Shaft is oval-shaped in cross-section. Articular facets are sharp and represent oblique ridges. The ulnar notch is prominent. Articular facet for lunate has a deep and narrow depression which is large and oval in *Bubalus*. Styloid process is sharp and much longer than in *Bubalus*.

REMARKS

This is one of the predominantly found genera in Pleistocene sediments of the Narmadā Valley, being represented by a large number of cranial, dental and post-cranial elements, often in a better state of preservation. Epiphyseal fusion and dental wear suggest the presence of predominantly an adult population with some numerically insignificant exceptions of younger individuals.

The phylogeny of the genus *Bos* has been a matter of controversy over the last hundred years or so. The ancestral history of *Bos* sp. was first discussed by Lydekker (1886) who suggested that *Bos namadicus* has evolved from either *Bibos* or *Bos gaurus*. However, Pilgrim (1939) was of the opinion that due to the major differences like frontal length, height of occipital crest and depth of temporal grooves, *Bibos* does not qualify as an

Pl. 5.1: NMD/1053/DC: Turtle, partial shell, carapace view.

Pl. 5.2: NMD/1053/DC: Turtle, partial shell, plastron view.

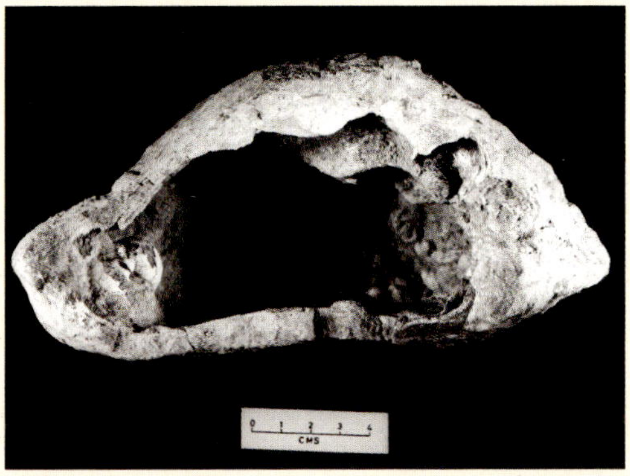

Pl. 5.3: NMD/1053/DC: Turtle partial shell, posterior view.

Pl. 5.4: NMD/1239/DC: A part of maxilla of *Crocodylus palaeindicus* (left: Dorsal view, right: Palatal view).

Pl. 5.5: NMD/1277/DC: Upper jaw of *Equus namadicus* (occlusal and labial views).

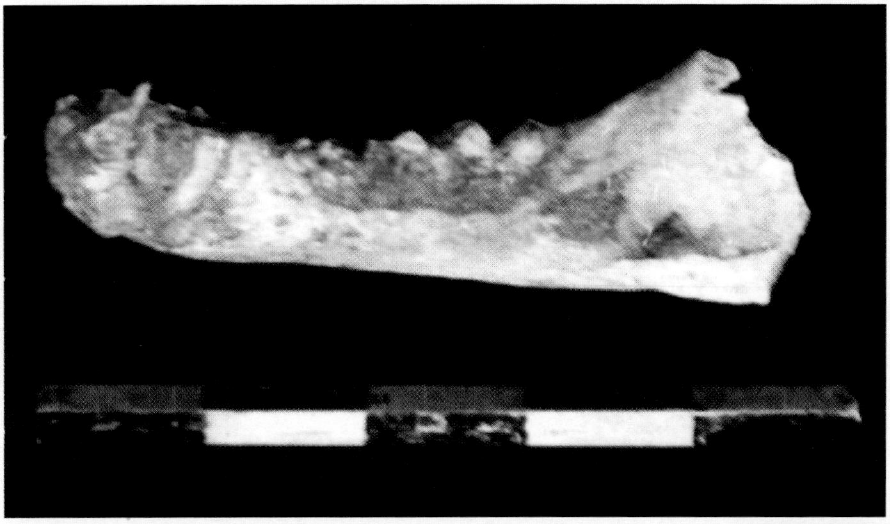

Pl. 5.6: Lower jaw of a carnivore.

Pl. 5.7: Part of a jaw of a carnivore.

Pl. 5.8: A damaged skull of *Bubalus palaeindicus*, dorsal view.

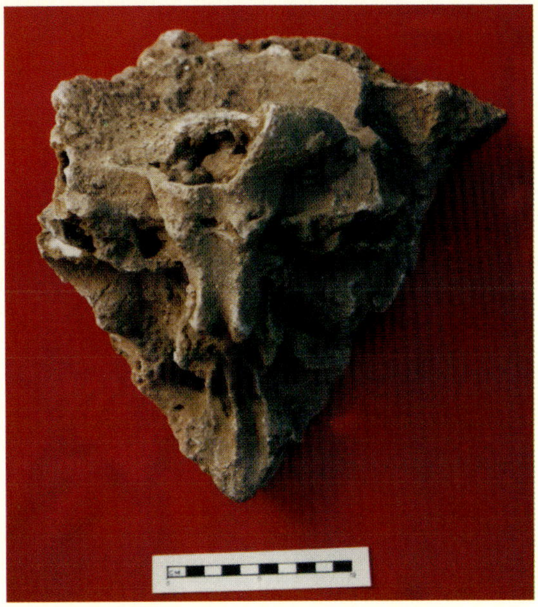

Pl. 5.9: A damaged skull of *Bubalus palaeindicus*, palatal view.

Pl. 5.10: NMD /1031/: Atlas vertebra of *Bubalus palaeindicus*.

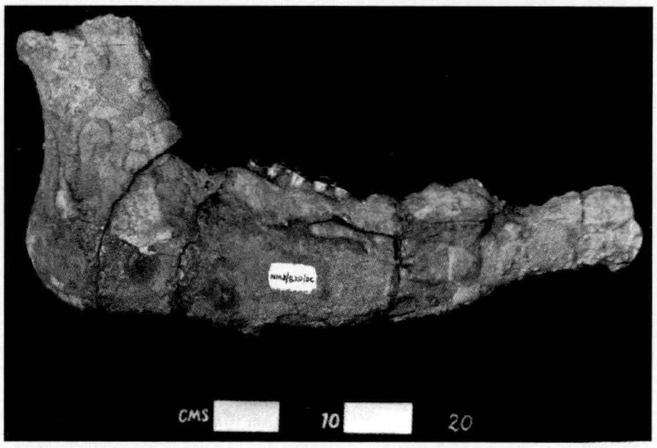

Pl. 5.11: NMD/820/DC: Right lower jaw of *Bubalus* n.sp., buccal view.

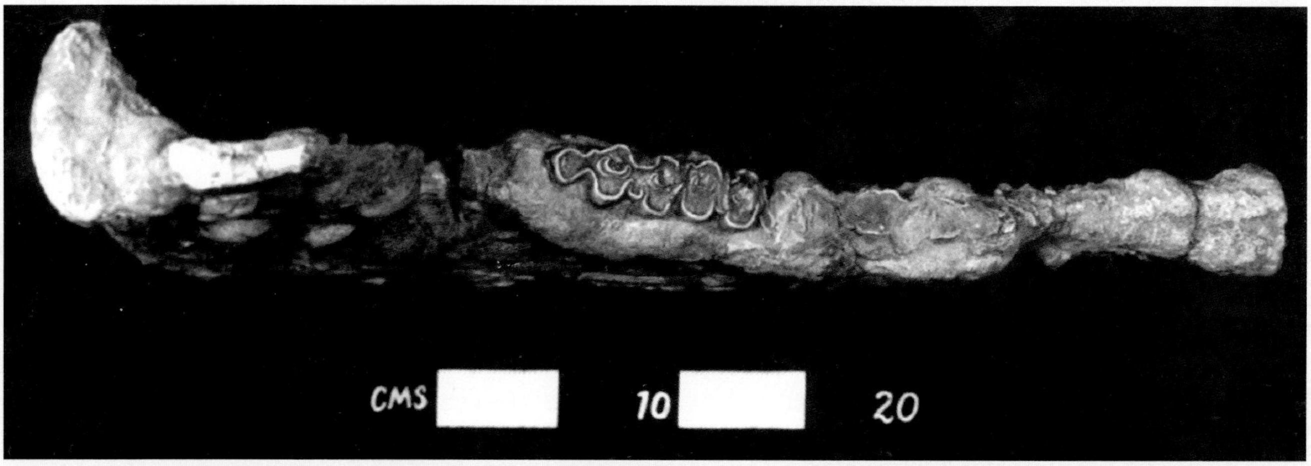

Pl. 5.12: NMD/820/DC: Right lower jaw of *Bubalus* n.sp., crown view.

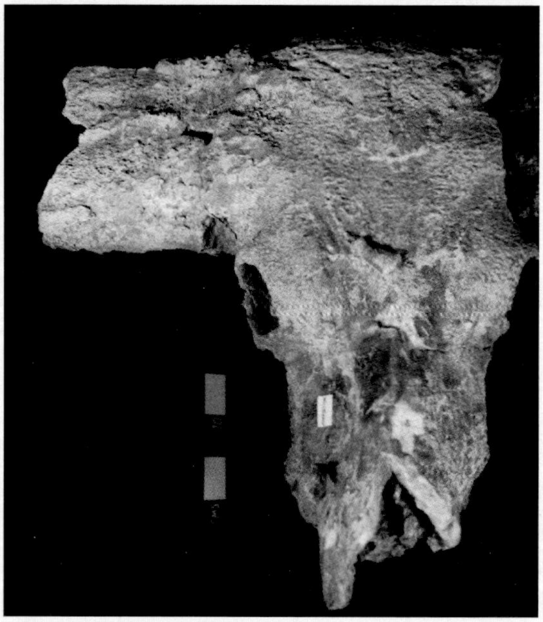

Pl. 5.13: NMD/819/DC: Skull with a broken horncore of *Bubalus* n.sp.

Pl. 5.14: NMD/1178/DC: A left lower jaw of *Bubalus palaeindicus*.

Pl. 5.15: DVK/1/94: Part of horncore of *Bubalus palaeindicus*.

Pl. 5.16: Parts of horncore of *Bubalus palaeindicus*.

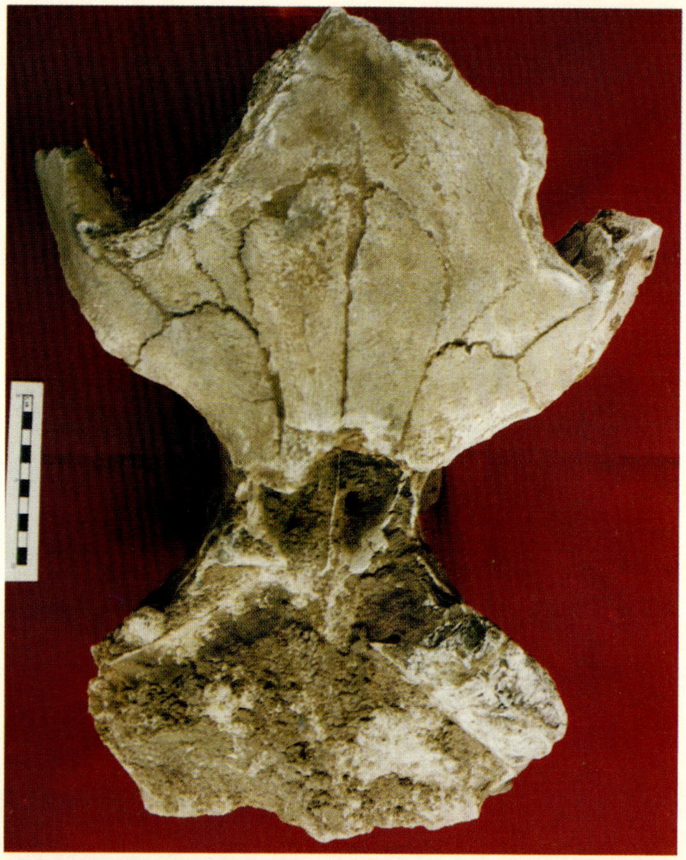

Pl. 5.17: Skull of *Hexaprotodon palaeindicus* (anterior part broken), dorsal view.

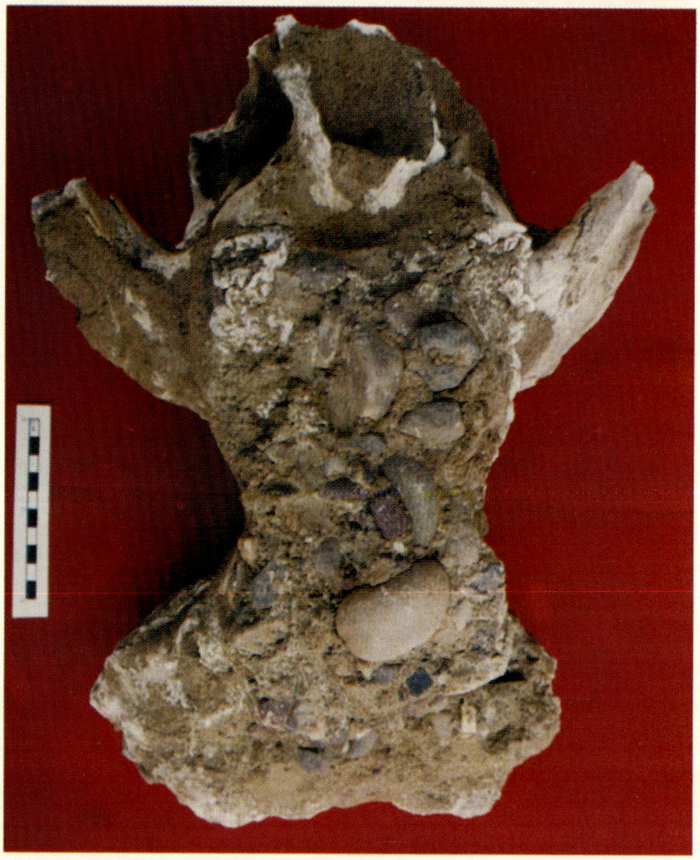

Pl. 5.18: Skull of *Hexaprotodon palaeindicus* (anterior part broken), palatal view.

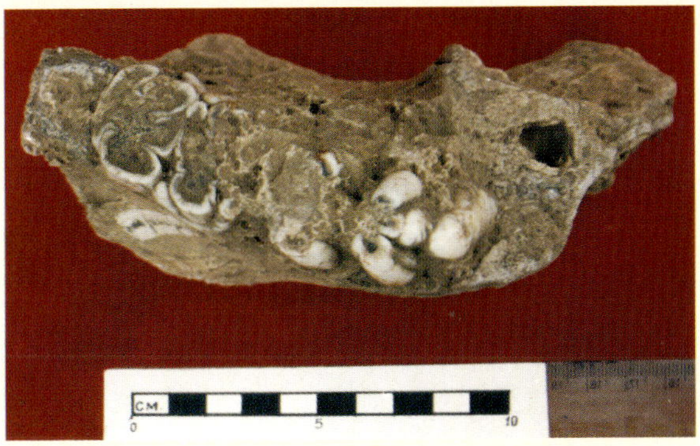

Pl. 5.19: Lower jaw of *Hexaprotodon* sp.

Pl. 5.20: NMD/1080/DC, 1081 and 1082: Canine fragments of *Hexaprotodon* sp.

Pl. 5.21: NMD/1102/DC: Posterior end of mandible of *Hexaprotodon* sp.

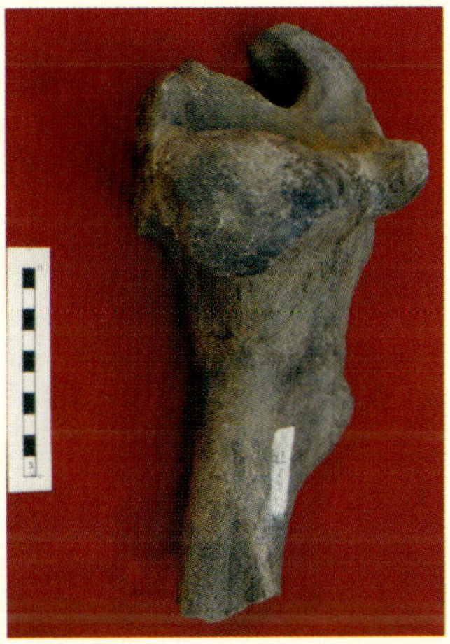

Pl. 5.22: NMD/1091/DC: Proximal humerus of *Hexaprotodon* sp.

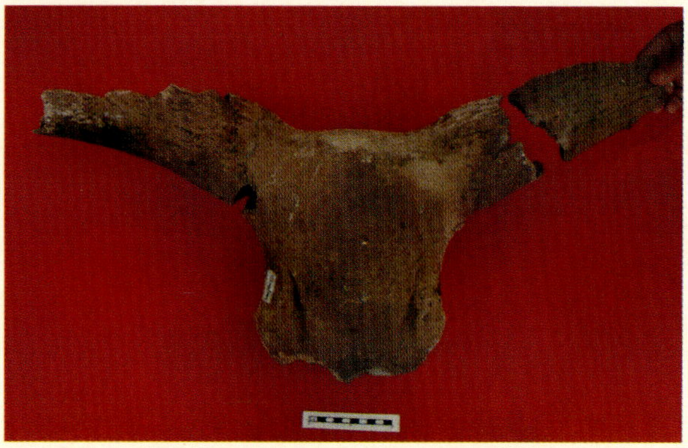

Pl. 5.23: NMD/1067/DC: Skull of *Bos namadicus*, dorsal view.

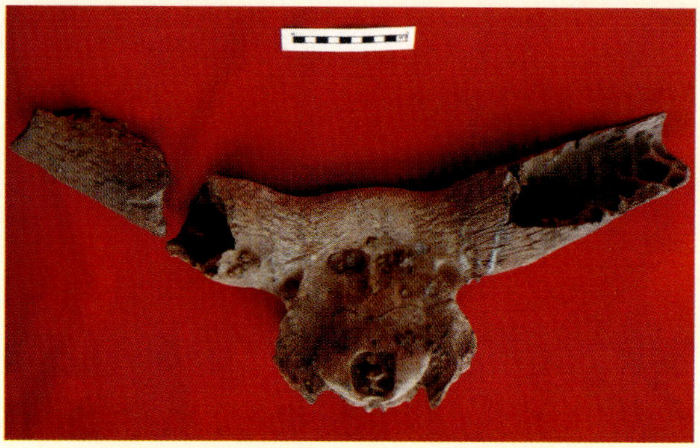

Pl. 5.24: NMD/1067/DC: Skull of *Bos namadicus*, occipital view.

Pl. 5.25: Skull of *Bos namadicus*, side view.

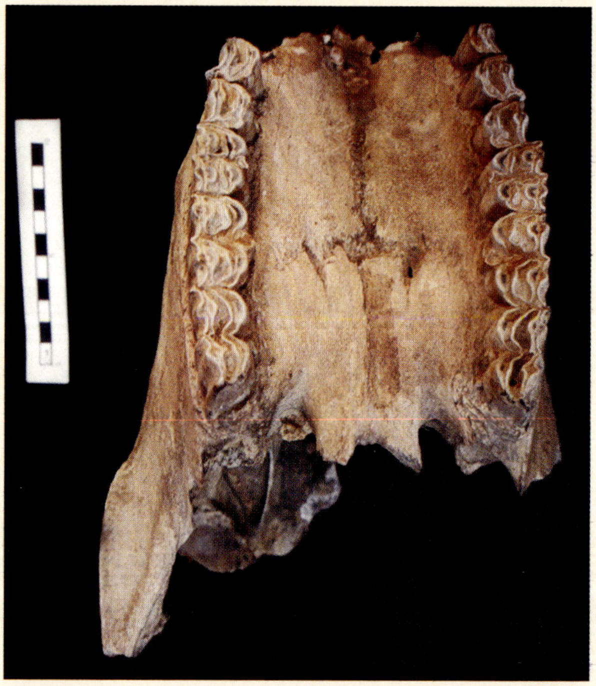

Pl. 5.26: Skull of *Bos namadicus*, palatal view.

Pl. 5.27: NMD/1185/DC (left) and NMD/1243/DC (right): Lower jaws of *Bos namadicus,* 2 parts.

Pl. 5.28: Horncores of *Bos namadicus,* 2 parts.

Pl. 5.29: Horncore part of a large Bovid.

Pl. 5.30: NMD/1141/ DC: Metacarpal distal end of *Bos namadicus.*

Pl. 5.31: NMD/1188/DC: Lower molar fragment of *Elephas*, side view.

Pl. 5.32: NMD/1188/DC: Lower molar fragment of *Elephas*, occlusal view.

Pl. 5.33: NMD/1099/DC: Lower molar of *Elephas*, side view.

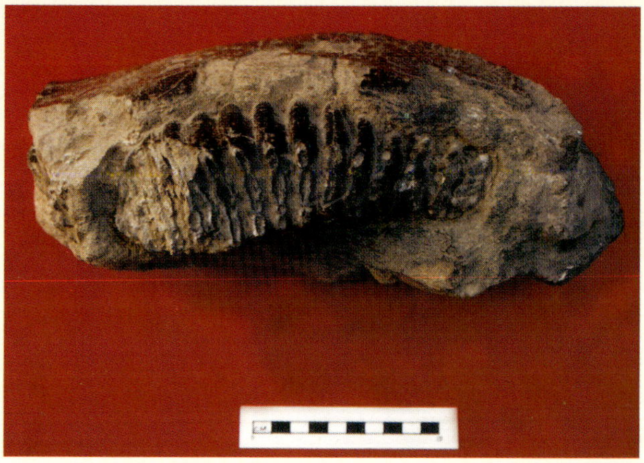

Pl. 5.34: NMD/1099/DC: Lower molar of *Elephas*, occlusal view.

Pl. 5.35: Lower jaw of *Elephas hysudricus* with a broken symphysis.

Pl. 5.36: Lower jaw of *Elephas hysudricus* with
a broken symphysis (this was actually part of *Pl.* 3.35).

Pl. 5.37: NMD/1107/DC: Lower molar of *Elephas*.

Pl. 5.38: NMD/1109/DC: Molar fragment of *Elephas*.

Pl. 5.39: Lower molar of *Elephas*, crown view.

Pl. 5.40: Lower molar of *Elephas*, side view.

Pl. 5.41: NMD/1123/DC: Tusk fragment of *Elephas,* front view.

Pl. 5.42: NMD/1123/DC: Tusk fragment of *Elephas,* side view.

Pl. 5.43: Femur head of *Elephas,* top view.

Pl. 5.44: Femur head of *Elephas,* inner view.

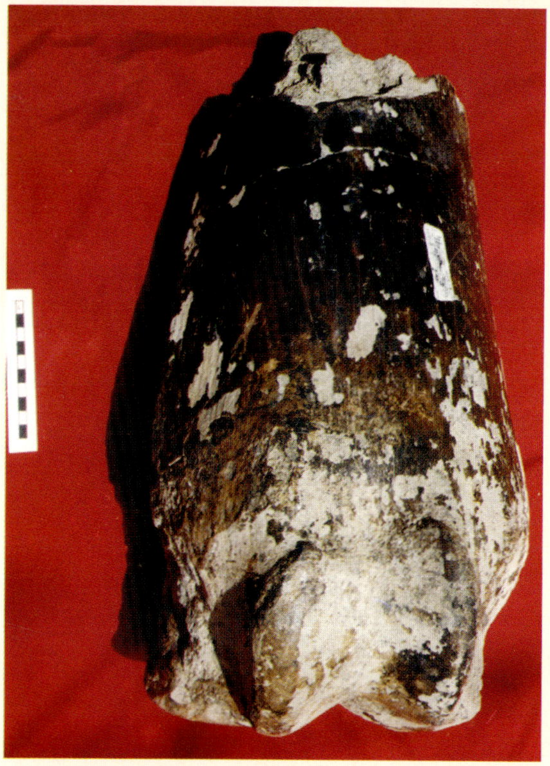

Pl. 5.45: NMD/1014/DC: Distal half of femur of *Elephas*, outer view.

Pl. 5.46: NMD/1014/DC: Distal half of femur of *Elephas*, inner view.

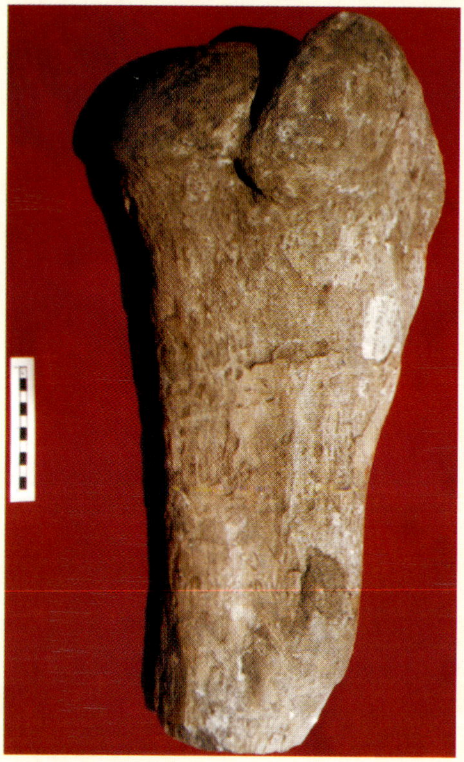

Pl. 5.47: NMD/1096/DC: proximal half of humerus of *Elephas*, outer view.

Pl. 5.48: NMD/1236/DC: proximal half of tibia of *Elephas*, outer view.

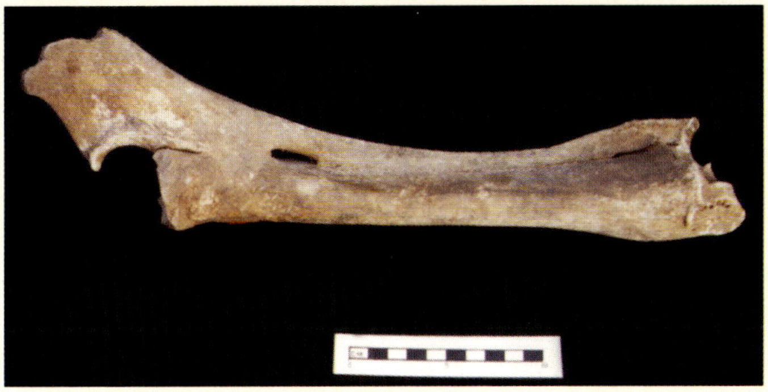

Pl. 5.49: Radio-ulna of a large mammal.

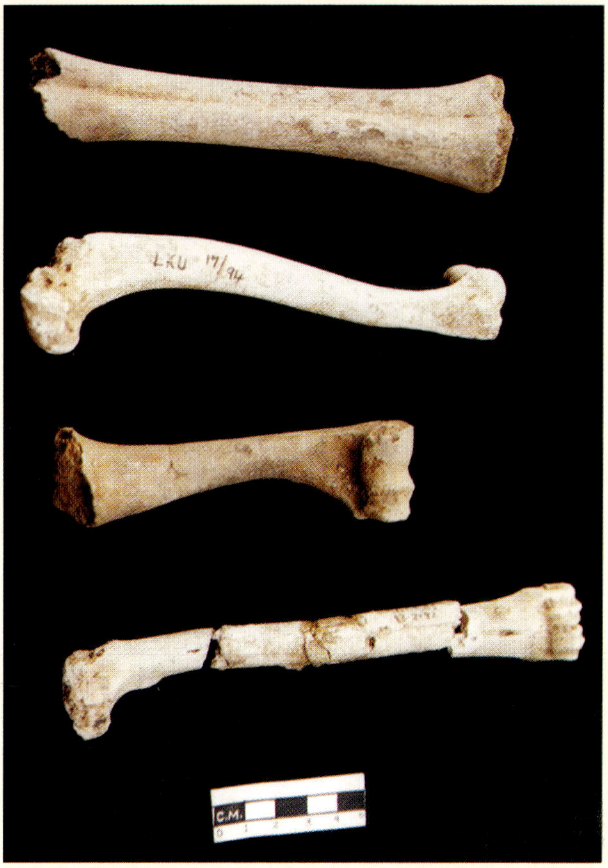

Pl. 5.50: Long bones of medium sized mammals (metapodials, humerus, femur), the first at bottom (NMD/1079/DC) is a metatarsal of *Antilope cervicapra* (bottom).

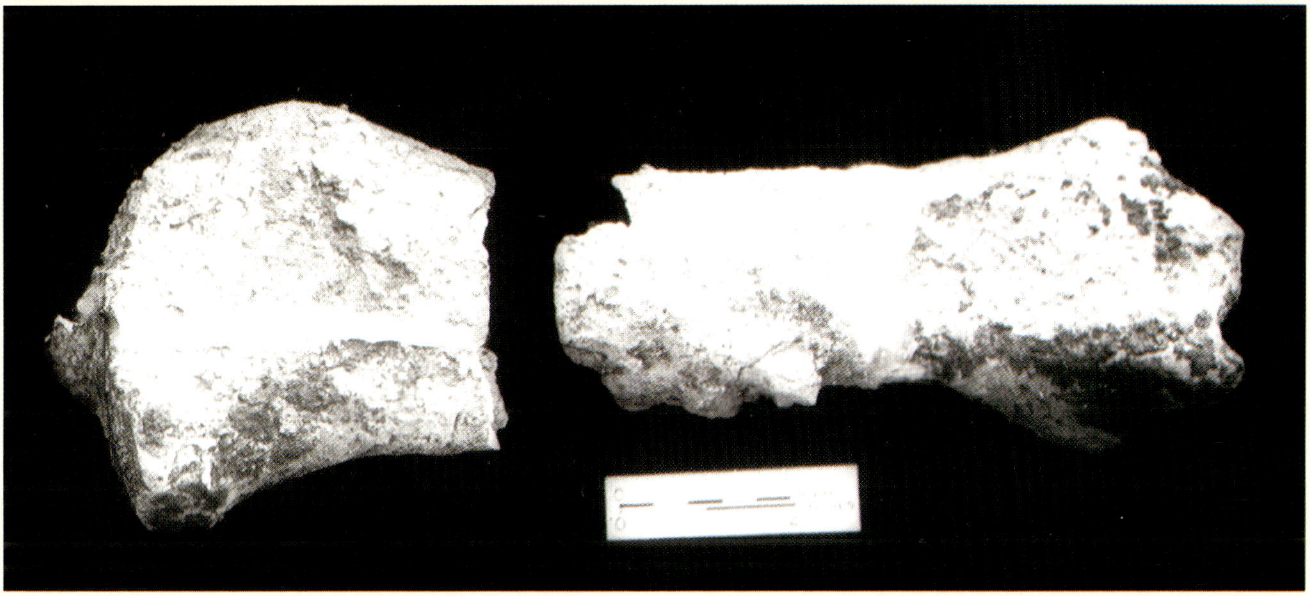

Pl. 5.51: Tibia of a bovid (broken).

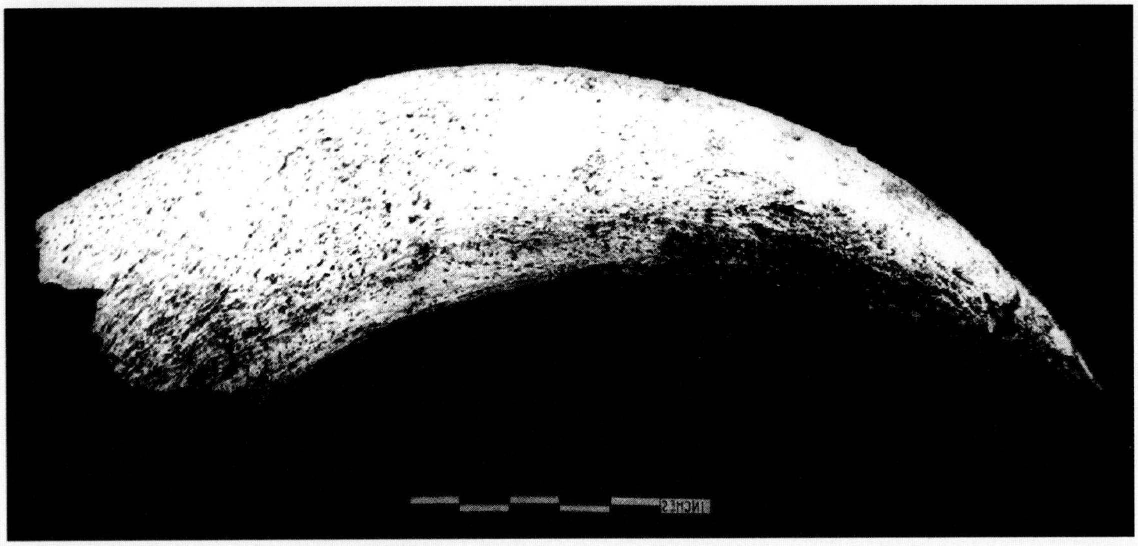

Pl. 5.52: Horncore of *Bos* sp.

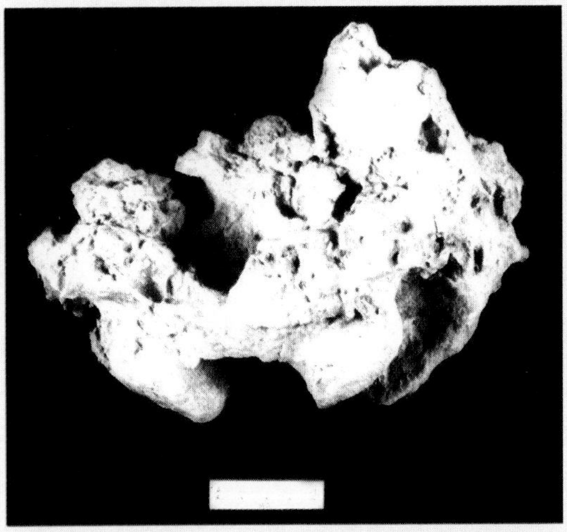

Pl. 5.53: Posterior part of skull of a bovid.

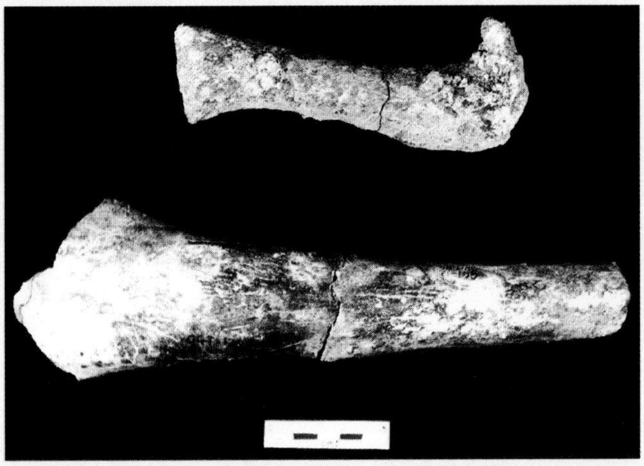

Pl. 5.54: Thoracic spine of a large mammal (lower).

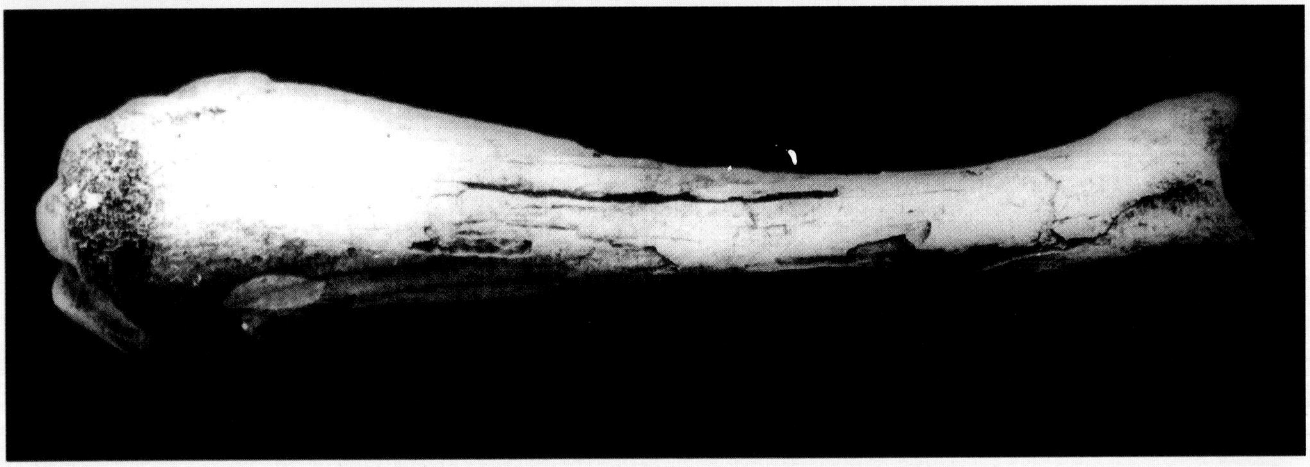

Pl. 5.55: A tibia with longitudinal cracks on the shaft.

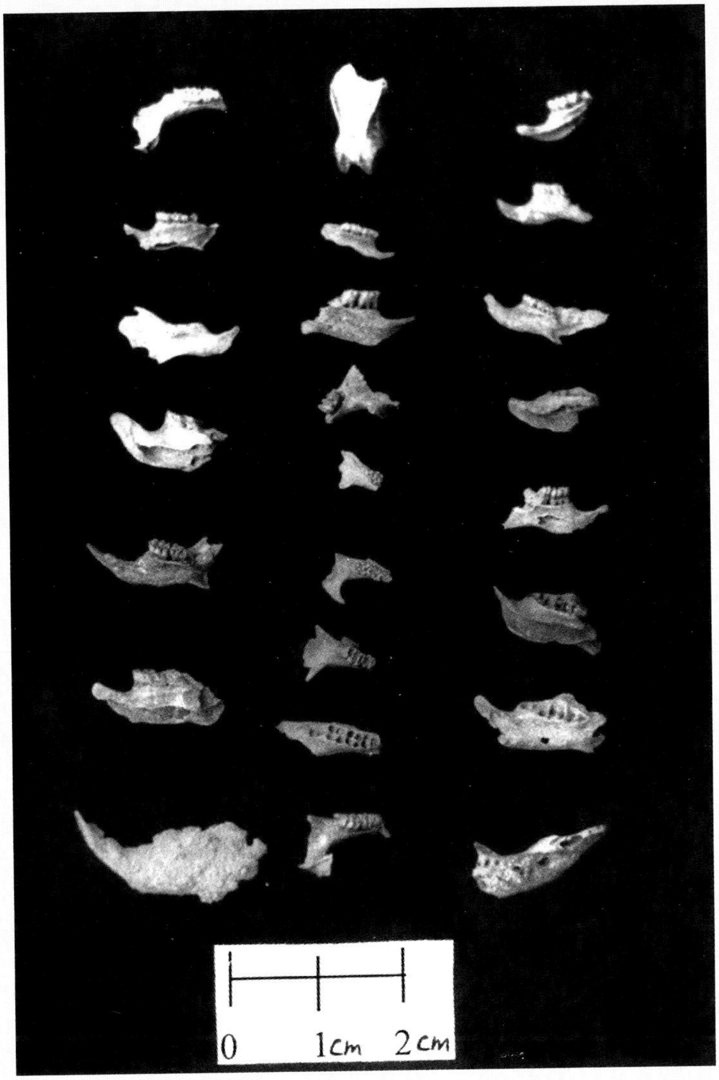

Pl. 5.56: Upper & lower jaws and a skull of rodents.

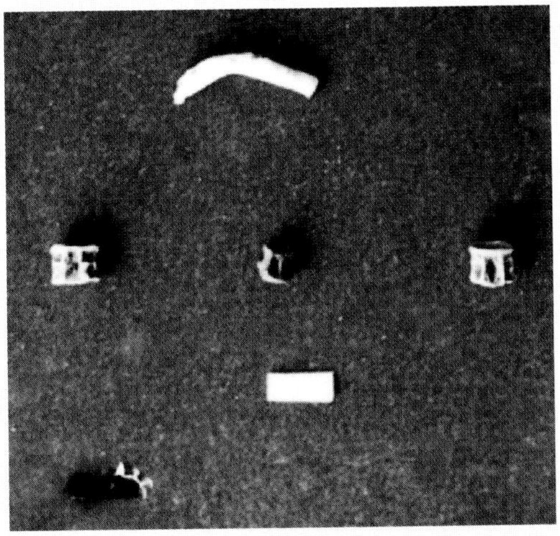

Pl. 5.57: Vertebrae of fish.

Pl. 5.58: A shelly gravel.

ancestor to *Bos namadicus.* He suggested that *B. namadicus* and *B. primigenius* have evolved from a common ancestor, viz. *Bos acutifrons* of the Śiwāliks of NW India. Recently, Epstein and Mason (1984) have advocated that *Bos namadicus* perhaps evolved through Asiatic variety of *Bos primigenius,* which initially originated from European *Bos primigenius.*

Bos namadicus has generally been considered as an ancestor of living cattle, *Bos indicus* (Badam, 1984; Grigson, 1985; Kumar, 1985). Grigson (1985) argued that probably *Bos indicus* originated in India, either through cross-breeding with European *Bos taurus* or by autochthonous domestication of *Bos namadicus* in the early Holocene times. Badam (1984) also suggested that *Bos indicus* might have originated through autochthonous domestication of *Bos namadicus.* However, minor morphological changes in the cranium and horns have been explained to be the result of process of domestication (Grigson, 1985). The cranial and dental metric and non-metric analyses of both the species have revealed that by and large the metrical observations on the skulls are slightly larger though the length of basioccipital and maxilla have remained consistent as in *Bos indicus* (Kumar, 1985). Metrical analysis of cheek teeth, however, does not provide any means of distinction.

Recently, Joglekar and Thomas (1990) discussed the ancestry of *Bos* species with the help of genetic evidence and biostatistical considerations and suggested that the status of *Bos namadicus* as a separate and distinct species from *Bos indicus* no longer holds valid. However, these observations need to be verified by numerically substantial cranial and dental material of *Bos namadicus* and *Bos indicus* from different geological provenance for any definite answer to the riddle of cattle phylogeny. At the present stage of our knowledge based on metrical and non-metrical observations it is strongly advocated that *Bos acutifrons, Bos namadicus* and *Bos indicus* form a continūm in India from Lower Pleistocene to Holocene.

Bubalus palaeindicus Falconer, 1868

SYSTEMATIC PALAEONTOLOGY

Class	:	Mammalia Linnaeus, 1758
Order	:	Artiodactyla Owen, 1848
Family	:	Bovidae Gray, 1821
Genus	:	*Bubalus* Linnaeus, 1758
Species	:	*Bubalus palaeindicus* Falconer, 1868

SYNONYMY

1868 *Bos palaeindicus* Falconer, *Pal. Mem.,* I, p. 280.

1878 *Bubalus palaeindicus*, Lydekker, *Pal. Ind.* (10) I, p. 132.

1885 *Bubalus buffelus* var. *palaeindicus*, Lydekker, *Cat. Foss. Mamm. Brit. Mus.*, p. 28.

1898 *Bos bubalus palaeindicus*, Lydekker, *Wild Oxen, Sheep and Goats*, p. 247.

1939 *Bubalus palaeindicus*. Pilgrim, *Pal. Ind.*, 36, pp. 256-57.

DIAGNOSIS

A *Bubalus* very close to the typical race of *Bubalus bubalis* but of larger size with a more convex forehead, horncores more divergent and mainly directed outward but with a slight upward curvature, inner keel more rounded, temporal fossae narrower at their hinder end and indenting the occipital surface less deeply; nasals more arched, palate more elongated backwardly; basicranial axis less inclined to facial axis (Pilgrim, 1939: 256).

TYPE

Skull, F 193, Geological Survey of India, Calcutta.

ADDITIONAL MATERIAL

NMD/1178/DC: Left ramus with M1 and M2	:	*Bubalus palaeindicus*
NMD/1182/DC: Left ramus, M3	:	*Bubalus palaeindicus*
NMD/1087/DC: Humerus: distal, left	:	*Bubalus palaeindicus*
NMD/1193/DC: Femur: proximal, left	:	*Bubalus palaeindicus*
NMD/1195/DC: Tibia: distal, right	:	*Bubalus palaeindicus*

UNDESCRIBED MATERIAL

NMD/1085/DC: Tibia: proximal left	:	*Bubalus palaeindicus*
NMD/1031/DC: Atlas vertebra	:	*Bubalus palaeindicus*
NMD/1124/DC: Femur: distal right	:	*Bubalus palaeindicus*
NMD/1166/DC: 2nd phalanx, distal	:	*Bubalus palaeindicus*

NMD/1178/DC: Left ramus with M1 and M2 (*Pl.* 5.14)

The specimen is represented by a freshly broken (post-fossilization) ramus with well preserved M1 and M2. Alveolus for P4 shows a fresh removal of the premolar. Alveolus for P3 is also seen.

M1

Inner cusps (meta- and entoconid) are oval-shaped. Outer cusps, viz. proto- and hypo-conid are U-shaped and they are linked by a loop like basal pillar. Pre- and post-fossettes are well developed and spread out mesio-distally. Median ribs are strong, valleys are deep.

M2

The characters of the tooth elements resemble those of M1. Enamel is thick and more crenulated than in M1. Pre- and post-fossettes are more elongated. Basal pillar is erupted close to the crown height but is not yet fused to the outer cusps. Median ribs are strong and valleys prominent.

NMD/1182/DC: Left ramus with M3

The specimen is represented by a larger aspect of ramus with alveoli of M1 and M2 and an intact M3. First two molars are broken at the alveolar edge of the jaw. The enamel of posterior, inner cusp of third molar is also damaged. Anterior inner cusp is intact and is of an oval shape. Both the buccal cusps are well preserved and have a well defined U-shape with thick enamel. Hypoconulid is large with a thick enamel fold at the posterior-most end of the tooth, pointing to its identification as *Bubalus*. Pre- and post-fossettes are elongated with crenulated enamel directed mesio-distally.

NMD/1087/DC: Humerus, distal left

The humerus is represented by distal end and a part of shaft which shows oblique fractured edge. Above the medial epicondyle, outer table of bone is chipped off. Medial epicondyle is laterally spread and lateral epicondyle is dorso-ventrally elongated. Olecranon fossa is relatively deep, oval in shape and antero-posteriorly compressed. Trochlea is expanded laterally. Coronoid fossa is moderately deep and latero-medially elongated. On the dorsal side of the specimen shaft is fractured up to coronoid fossa.

NMD/1193/DC: Femur, Proximal left

The specimen is represented by a well preserved part of shaft and proximal end of femur. Shaft has a somewhat triangular outline as seen in the cross-section. There is a partial deep crack running from shaft to the outer side of trochanteric fossa which appears to be a pre-burial phenomenon. Proximal epiphyseal cap indicates a recently fused condition. Trochanteric fossa is deep. Proximal end of the trochanter major is somewhat eroded while the head is intact.

REMARKS

Bubalus bubalis, the sole survivor of the genus *Bubalus* in India, is believed to have descended from the Pleistocene species *Bubalus palaeindicus* through a wild prototype *Bubalus arnee* (which lived formerly in India and South Asia) and which presently inhabits Nepal and East Asia. *Bubalus bubalis* is represented by both wild and domestic forms which, however, osteologically cannot be seperated from each other (Clutton-Brock, 1965; Badam, 1984).

The genus *Bubalus* has been represented by various species, found in sediments of Upper Śiwāliks of the NW and the Peninsular river valleys in India. However, only two species are taxonomically determinable, viz. *Bubalus platyceros* and *Bubalus palaeindicus.* The former is characterized by flattened and narrow forehead, nasals far back, horncores less divergent with a typical triangular cross-section and with three sharp keels; parietal large, open temporal fossa and occipital low in proportion to its breadth (Pilgrim, 1939).

Bubalus palaeindicus (reported from Pleistocene deposits of Peninsular India and Upper Śiwāliks of NW India) has convex forehead, divergent horncores with inner keel more rounded, narrow temporal fossae indenting the occipital surface shallow, nasal arched, palate prolonged and basicranial axis less inclined to facial axis (Pilgrim, 1939: 256).

Recently, an unusually large-sized complete cranium, lower jaws and vertebrae of *Bubalus* n. sp. (*Pl.* 5.11 to *Pl.* 5.13) have been discovered *in situ* on the left bank of Narmadā at Barmanghāṭ. Cranial morphological features that differ in characters from *B. palaeindicus* are:

(i) palate relatively contracted unlike in *palaeindicus,*

(ii) maxillary teeth hypsodont but little broader than in *palaeindicus,*

(iii) frontal eminence convex and at a right angle to the basicranium (which in *palaeindicus* forms a slope).

Cranial and horncore measurements suggest that this is perhaps the largest buffalo ever found in Pleistocene deposits of South Asia. Further work on this specimen is in progress.

Being a closely related genus there are only a few non-metric morphological features, given below, which help to differentiate *Bubalus* from *Bos*:

(i) Rib-like flattened horncores.

(ii) Coronoid process extending beyond mandibular condyle.

(iii) Enamel folds at the distal ends in lower and upper M3 oval-shaped and very prominent.

(iv) The groove between lateral and medial tuberosities in the proximal part of humerus shallow than in *Bos.* Olecranon fossa very deep in *Bos.*

(v) The space of intercondyloid fossa in tibia wider than in *Bos.*

(vi) Intercondyloid fossa of femur wider than in *Bos.*

(vii) Distal ends of metacarpals and metatarsals exhibit sideward expansion, particulary just above the trochlea. This denotes adaptation to facilitate walking in mud (Hooijer, 1958).

Bos/Bubalus

There are a few dental and post-cranial elements which, owing to the lack of diagnostic features, are being described independently under the heading Bos/Bubalus.

NMD/1185/DC: Left mandible with P4 - M3 (*fig.* 5.2)

The mandible has preserved only first and second molars. P4 as well as M3 have lost their anterior and posterior ends and even the jaw at M3 is abruptly broken. Teeth are hypsodont and their elements are well defined.

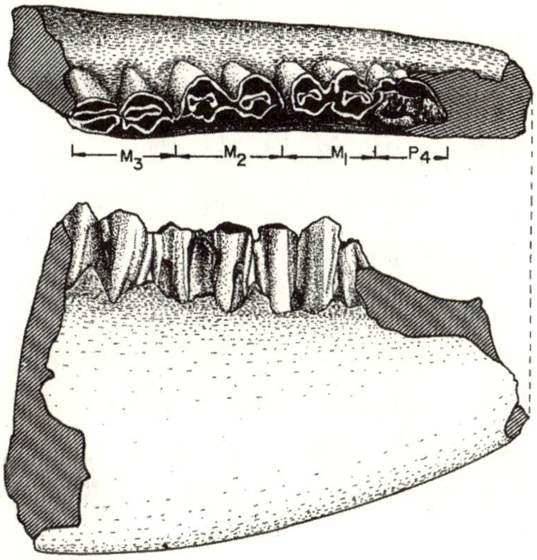

fig. 5.2: NMD/1185/DC: Left mandible of Bos/Bubalus

M1

Except for the anterior enamel fold of the anterior group of cusps, rest of the tooth is complete. Inner cusps are oval and are seperated by pre- and post-fossettes. Proto- and hypoconid (outer cusps) are V-shaped and directed at a right angle to the jaw axis. Pre- and post-fossettes are small with less crenulated enamel and are proportionate to other elements of the tooth. Basal pillar is erupted upto the crown height. Median ribs are strong and valleys are deep.

M2

Inner cusps (meta- and entoconid) are damaged but outer ones (proto- and hypoconid) are well preserved and V-shaped. Pre- and post-fossettes are more elongated and crenulated. Basal pillar is well developed. Median ribs are strong and valleys are partially filled with matrix.

M3

This is a poorly preserved tooth. Hypoconulid is absent, dental elements are not yet fully developed. Fossettes are thin and elongated. Both inner and outer cusps are sharp, V-shaped and restricted by narrow enamel outline.

NMD/1078/DC: Right ramus with M1 & M2

The ramus is poorly preserved with M1 and erupting M2. Broken edges of the ramus are slightly abraided. Surface of ramus shows some superficial mosaic cracks.

M1

The tooth wear suggests a young age for the individual. Inner and outer cusps are not developed. Pre- and post-fossettes are hollow and only the former preserves enamel outline. Median ribs are weak. Basal pillar is developed up to half of the crown height. M2 is in erupting stage.

NMD/1183/DC: Right ramus with M2 & M3

The specimen is represented by a well preserved ramus with high crowned teeth. Enamel of the buccal cusps is damaged.

M2

Crown view shows oval inner cusp (ento- and metaconid) and V-shaped outer ones (proto-and hypoconid). Pre- and post-fossettes are mesio-distally elongated and enamel is less crenulated. Basal pillar has erupted upto the crown height. Median ribs are quite strong and valleys are filled with matrix.

M3

Enamel of this tooth is slightly eroded. Inner cones are oval and outer ones are V-shaped. Pre- and post-fossettes are elongated with thick enamel. Hypoconulid is thick, with a bigger loop directed bucco-lingually and a small pinch towards the distal end of the jaw.

NMD/1186/DC: Right ramus with P4 & M1

The specimen is a poorly preserved ramus with broken crown of P3, complete P4 and slightly damaged M1. Crown view of the fourth premolar represents thick and deep folded enamel which is mildly worn out. Crown of the first molar is slightly eroded and enamel folds at the lingual side are broken. Pre- and post-fossettes are small and enamel is less crenulated. Basal pillar is erupted up to the crown height. Broken ends of the ramus are abraided and there are surficial striations running parallel to the horizontal ramus.

NMD/1184/DC: Left ramus with M2

The specimen is ill preserved. Second molar is high crowned. Inner and outer cusps are poorly developed. Fossettes are hollow and basal pillar is erupted up to the half of crown height.

NMD/1177/DC: Upper DM1/DM2

The specimen is a poorly preserved maxillary fragment with deciduous molars which are fragile.

DM1

Tooth is broad and rectangular. Inner and outer cusps are least developed and fossettes are represented by deep fissures. Median ribs are moderately developed.

DM2

This is also weakly developed though the size is larger than DM1. Basal pillar is erupted up to half of the crown height.

NMD/1191/DC: Cervical Vertebra (*fig.* 5.3)

The specimen is represented by a partially preserved cervical vertebra. One of the transverse processes is broken and so is the neural spine at the dorsal end. Ventral aspect reveals a concave centrum with a widespread mammalliary process. Transverse foramen are filled with matrix. Neural arch is circular. Posterior zygapophysis are a pair of obliquely oriented

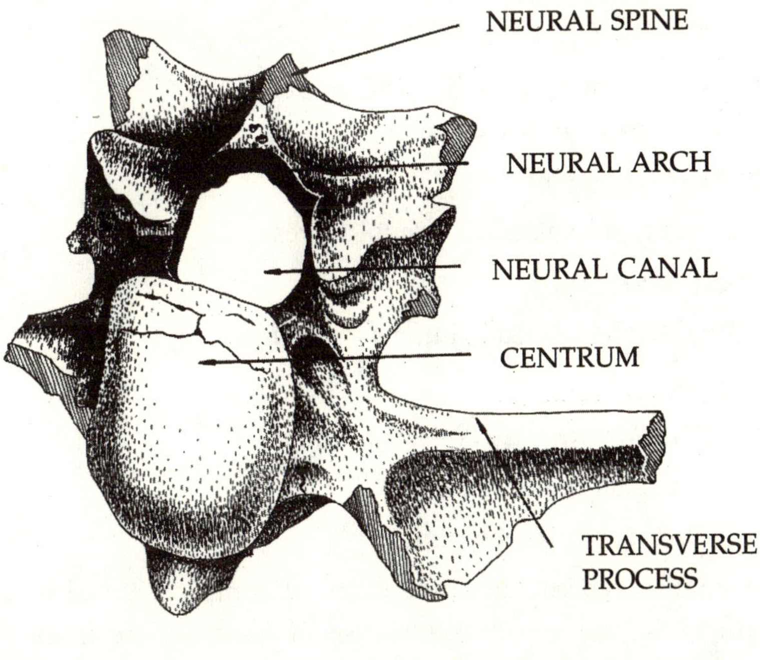

fig. 5.3: NMD/1191/DC: Cervical vertebra of *Bos/Bubalus*

flat processes which accommodate facets for ribs. Anterior zygapophysis have a similar arrangement though much on a higher elevation as seen from the cranial view.

***Antilope cervicapra* Linnaeus, 1758**

SYSTEMATIC PALAEONTOLOGY

Class	:	Mammalia Linnaeus, 1758
Order	:	Artiodactyla Owen, 1848
Family	:	Bovidae Gray, 1821
Genus	:	*Antilope* Pallas, 1776
Species	:	*Antilope cervicapra* Linnaeus, 1758

SYNONYMY

1758 *Antilope cervicapra* Linnaeus, *Systema Naturae.*

1865 *Antilope cervicapra,* Lydekker, *Cat. Foss. Mam. Brit. Mus. (Nat. Hist.)* II, 52.

1939 *Antilope cervicapra,* Pilgrim, *Pal. Ind.,* 26(1), 39.

DIAGNOSIS

Antelope with horns of considerable length and greatly twisted, making one revolution in a length of about 105 mm, cross-section of horncore circular, keels absent or obscure; no basal pillars in the lower molars (Pilgrim, 1939: 39).

TYPE

F100: A horncore fragment and F 287: upper M2.
Geological Survey of India, Calcutta.

LOCALITY AND HORIZON

Older Alluvium, Jamunā and Kurnool Caves.

ADDITIONAL MATERIAL

NMD/1079/DC: Metatarsal: *Antilope cervicapra*

DESCRIPTION

NMD/1079/DC: Metatarsal (*Pl.* 5.50 : bottom)

Locality: Bheḍaghāt (palaeochannel)

Geological Unit: Sandy pebbly gravel

The specimen is represented by an almost complete tarsal bone. The proximal end is, however, broken. Edges of specimen are fresh. Shaft is cylindrical and cross-section is triangular. Foramen towards the distal trochlea is vertically elongated. There is a pair of pully-like trochlea which are small and laterally compressed.

REMARKS

Antelope has been reported from various geological deposits of the Pleistocene being represented by *Antilope subtorta, Antilope planicornis* and *Antilope cervicapra* of which the last one appears for the first time in Upper Pleistocene. *Antilope cervicapra* continued till the Holocene and is reported from most of the archaeological sites in India.

The present collection consists of a very small number of bones (two metapodials). The mode of preservation is excellent and both the specimens have come from different age groups, e.g. juvenile and sub-adult which is indicated by the state of epiphyseal fusion.

As regards its ancestral history, Pilgrim (1939) suggested that the species may have descended from the genus *Helicotraqus* of Pontian. The Śiwālik *A. subtorta* had not developed present day characteristics of antelope, e.g. the horncores were less twisted with circular cross-section and the keel was completely absent. Unlike living antelopes, mandibular teeth possessed basal pillars. Species appearing later in the process of antelope evolution, viz. *Antilope planicornis* shows gradual appearance of diagnostic features, such as, the horncores were less twisted but compressed and possessed a weak postero-external keel. However, basal pillars were still seen in lower molars. By the beginning of the Upper Pleistocene the genus *Antilope* had attained all the diagnostic skeletal morphological features of the living antelope. This is supported by absence of keel; horncores with circular cross-section and disappearance of basal pillars in the lower teeth of *Antilope cervicapra.* The dentition in antelope is selenodont in character. The lingual enamel of the teeth is U-shaped which differentiates it from *Ovis/Capra* where it is V-shaped.

Ovis/Capra

SYSTEMATIC PALAEONTOLOGY

Class : Mammalia Linnaeus, 1758

Order : Artiodactyla Owen, 1848

Family : Bovidae Gray, 1821

Genus : *Ovis/Capra* Linnaeus, 1758

NMD/1197/DC: Right Humerus

The specimen is a well preserved humerus except for the proximal epiphyseal end and the tip of the articular surface of the medial epicondyle.

Shaft is rounded and curves towards distal end and merges into a shallow coronoid fossa posteriorly. Olecranon fossa is triangular of which the line of lesser angle is directed towards the lateral epicondyle. The articular surface of the medial epicondylus reaches half-way up the posterior articular surface of the trochlea (which in *Capra* reaches nearly the poximal end of it). The medial and lateral epicondyles, capitulum and trochlea are all well preserved.

Elephas hysudricus **Falconer and Cautley, 1846**

SYSTEMATIC PALAEONTOLOGY

Class : Mammalia Linnaeus, 1758

Order : Proboscidea Illiger, 1811

Family : Elephantidae Gay, 1821

Genus : *Elephas* Linnaeus, 1758

Species : *E. hysudricus* Falconer and Cautley, 1846

SYNONMY

1846 *Elephas hysudricus* Falconer and Cautley, *Fauna Antiqua Sivalensis*, Part-1.

1868 *Elephas hysudricus,* Falconer and Cautley, *Pal Mem.*, 1.

1885 *Elephas hysudricus,* Lydekker, *Cat. Siw. Vert. Ind. Mus.*, 1, 74.

1910 *Elephas hysudricus,* Pilgrim, *Rec. Geol. Surv. Ind.*, 40, 200.

1935 *Hypselephas hysudricus,* Colbert, *Trans. Amer, Phil. Soc.*, N.S., 26 33.

1942 *Hypselephas hysudricus,* Osborn, *Proboscidea*, 2, 1340.

1955 *Elephas hysudricus,* Hooijer, *Zool. Verhand*, 28, 107-10

DIAGNOSIS

Cranium with vaulted vertex, deeply concave frontals, orbits situated in a lower position; lamellar frequency less than in living elephants (6-8-8.5), height of the plates less than in the living elephants and the profile slightly S-shaped; marked difference between orbital and the premaxillary width; lower border of infraorbital foramen lost, skull much narrow at the temporal contraction (Osborn, 1942; Hooijer, 1955).

TYPE

British Museum No. M 3062, Upper second molar

ADDITIONAL MATERIAL

NMD/1179/DC: Upper molar: *Elephas hysudricus*

NMD/1188/DC: Lower molar: *Elephas hysudricus*

NMD/1179/DC: Upper molar

The molar is broken at the posterior end. It has preserved a total of seven lophs of which the first two have unfused cones. They represent 4 to 3 cones respectively. Lophs are slightly concave at roots. Enamel in other lophs is less crenulated. Lophodont sinus is moderately present. Interlamellar valleys are narrow and shallow.

NMD/1188/DC: Lower Molar (*Pl.* 5.31 & 5.32; *fig.* 5.4)

The molar is broken both posteriorly as well as anteriorly. The preserved

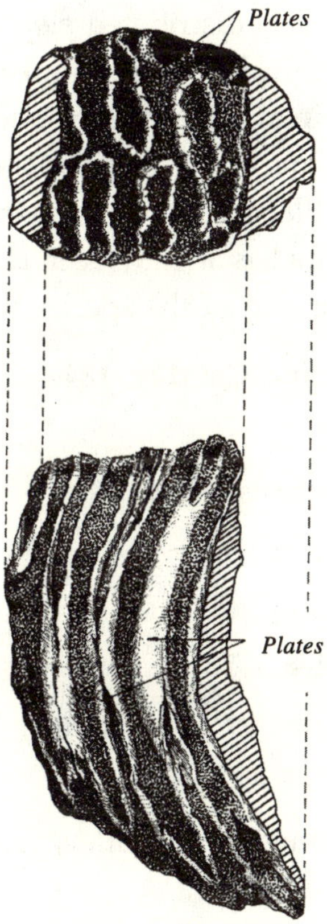

fig. 5.4: NMD/1188/DC Lower molar of *Elephas* (side and occlusal views)

lophs are worn out and their cones are unfused. The lophs are slightly concave anteriorly and convex at roots. Inter-loph valleys are narrow. Enamel is moderately crenulated. Lophodont sinus is rudimentary. Number of plates present is three.

REMARKS

Appearance of the genus *Elephas* along with *Equus* and *Cervus* is considered to coincide with the advent of the Pleistocene at 2.48 mys; Matuyama Magnetic epoch (Azzaroli and Napoleone, 1982; Badam, 1988).

The evolutionary history of *E. hysudricus* has generally been related to *Elephas planifrons* of Tatrot and Piñjor formations of Upper Śiwāliks (Gaur, 1981). The phylogenetic relationship between *E. hysudricus* and the living elephant (*E. maximus*) is uncertain, though cranially they resemble each other. The teeth of former are quite primitive than those of the latter. The cranium of *E. hysudricus* varies greatly from *Palaeoloxodon namadicus* and *P. africana*. Taking into consideration morphological parameters, *E. hysudricus* and *Palaeoloxodon namadicus* do not set conditions for the direct ancestry of the living Asian elephant (Hooijer, 1955).

This is one of the most widely distributed elephant species that lived during the Pleistocene period in India. In Peninsular river valleys the fossilized remains of this animal have been found in association with *Stegodon insignis ganesa*. Outside India, the fossils of *E. hysudricus* have been found from the Lower Pleistocene deposits of Upper Irrawady of Burma. A great deal of literature has already been published bearing upon its phylogeny and interrelationship of the species with other forms.

Elephas namadicus Falconer and Cautley, 1846

SYSTEMATIC PALAEONTOLOGY

Class	:	Mammalia Linnaeus, 1758
Order	:	Proboscidea Illiger, 1811
Family	:	Elephantidae Gray, 1821
Genus	:	*Elephas* Linnaeus, 1758
Species	:	*E. namadicus* Falconer and Cautley, 1846

SYNONYMY

1846-47 *Elephas namadicus* Falconer and Cautley, *Fauna Antiqua Sivalensis*, 1-7.

DIAGNOSIS

Ridge plates parallel and thin, ridge plate formula from twelve to nineteen in third molar, loxodont sinus absent or rudimentary, plicate enamel; broad rugose parieto-frontal crest, pre-maxillaries broad inferiorly; tusks relatively straight, slightly upcurved and incurved (Osborn, 1942).

TYPE

Elephas namadicus Falconer and Cautley, 1846

ADDITIONAL MATERIAL

NMD/1099/DC: Right lower molar

DESCRIPTION

The specimen is a lower molar broken on mesial and distal ends. As a result four of the lamellae are severely damaged. A total of eleven lamellae are present of which three are still erupting. Enamel is thin and less plicated, interlamellar spaces which are straight are filled with cement, loxodont sinus is rudimentary. Broken edges of the ramus are abraided. Superficial cracks can be seen on the surface of the ramus.

REMARKS

Loxodonta, Mammuthus and *Elephas* are three genera of the family *Elephantidae* of the Pleistocene period which have been differentiated on the basis of cranial characters. All three species have descended from the African genus of *Primelephas* which belongs to a primitive elephantine stock. The family *Elephantidae* has been divided into two sub-families, viz. *Stegotetrabelodontinae* and *Elephantinae*. Earlier, *Stegolophodon* sp. was believed by some scholars to be ancestral to all the elephants which, however, has been questioned on the grounds of insufficient morphological and stratigraphical data. Considering the affinity in terms of morphological parameters and correct geological provenance, *Primelephas gomphotheroides* (Lower Pliocene of Africa) may be considered a basal genus, which in turn gave rise to the later elephants both in Africa as well as in Asia.

Elephas namadicus is considered to have appeared at the time of extinction of *Stegodon*. However, its remains have been found contemporaneously with the latter forms (which were on their way out), from the mid-late Pleistocene deposits in India. Generally, on stratigraphical and morphological grounds *Elephas namadicus* and *Elephas antiquus* (of Europe) are considered contemporary (Pilgrim, 1905).

6

Fossils and Taphonomy*

THIS chapter deals with brief descriptions and taphonomic attributes of some Narmadā fossils which have been helpful in a derivation of palaeoecology and the factors leading to preservation of fossils. The taxonomic and taphonomic descriptions of the fossils studied are given below:

Specimen no. NMD/1053/DC

Locality	:	Gurāriāghāṭ
Horizon	:	Sandy pebbly gravel
Identification	:	Partial shell of a turtle

SYSTEMATIC PALAEONTOLOGY

Phylum	:	Vertebrata
Class	:	Reptilia
Order	:	Chelonia
Family	:	Emydidae Gray, 1825

DESCRIPTION (*fig.* 6.1)

The shell is of medium size and plano-convex in cross-section. It is represented by the posterior half with the cavity for tail at its posterior end. Two lateral cavities on either side of the cavity for tail are those of the hind limbs. The carapace is elevated towards the nuchal ridge and sloping towards the margins. The nuchal ridge is prominent and so are the impressions of a couple of nuchal plates which are roughly triangular with somewhat rounded sides. A few partial impressions of the pleural plates are also visible which are of hexagonal shape. The rest are however concealed in a cement coating which has enveloped a major portion of the shell. The plastron is roughly flat and smooth with a slight depression at its central part and tapering towards the edges to meet the marginal plates of the carapace. The dorsal and ventral surface of the shell is smooth.

---

* There may be some repetition of fossils as a few have already been referred to in chapter 5. However, the angles of study are somewhat different.

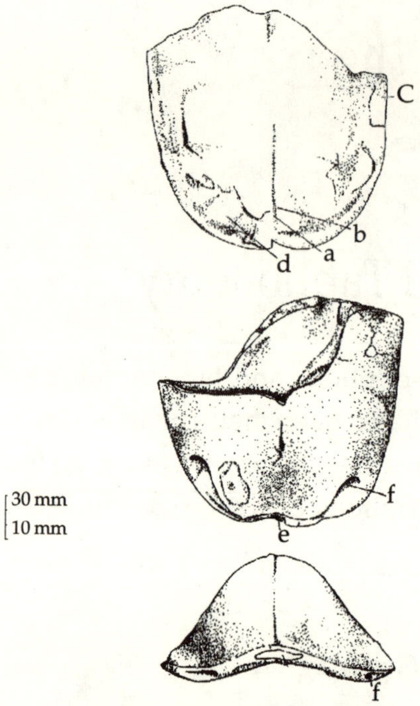

fig. 6.1: NMD/1053/DC Turtle partial shell

The specimen has not been identified specifically as it shows several mixed characteristics but is similar to the contemporary turtles found in the present-day rivers and resembles *Geoclemys sivalensis* of the Upper Śiwāliks of India (Tewari and Badam, 1969) and to *Geoclemys hamiltoni* found in the rivers of Pakistan by Minton (1966). This indicates the slow evolutionary process of this reptilian genus which has not changed much since the Pleistocene times.

MEASUREMENTS

Maximum height of the shell (central part)	:	96.8 mm
Maximum length of the shell (nuchal region)	:	189.0 mm
Maximum breadth of the carapace	:	195.0 mm
Thickness of the carapace (broken end)	:	12.1 mm
Thickness of the plastron	:	7.5 mm
Breadth of the cavity for tail	:	39.6 mm
Height of the cavity for tail	:	8.9 mm
Distance between the two cavities for hind limbs	:	105.5 mm

TAPHONOMIC ATTRIBUTES

The specimen is a partial shell of turtle with the posterior part intact. The entire specimen is enclosed in a cementing matrix except for some areas where the original shell cover can be seen. The shell is cracked irregularly all over which is a result of postmortem handling. The shell is associated with the silty clayey matrix of the Upper Narmadā Group which is characterized by fill deposits facilitating better preservation of animal remains. In the present case, the specimen must have got exposed and drifted from its place of origin resulting in the breakage as indicated by the freshness of the break. Importantly, a few gastropod and lamellibranch shells are attached with the matrix in the cavity which indicates its fresh water riverine origin. Similar shells were also collected in an earlier excavation which were dated to 31,150 ± 1820/1625 BP (Agarwal and Kusumgar, 1975), placing it securely in Upper Pleistocene time bracket.

Specimen no. NMD/1239/DC

Locality	:	Devakachār area
Horizon	:	Red Brownish Concretionary silt
Identification	:	Part of maxilla

SYSTEMATIC PALAEONTOLOGY

Phylum	:	Vertebrata
Class	:	Reptilia
Order	:	Crocodylia Gmelin, 1788
Family	:	Crocodilidac Cuvier, 1807
Genus	:	*Crocodylus* Laurenti, 1768
Species	:	*C. palaeindicus* Falconer, 1859

DESCRIPTION (*fig.* 6.2)

The specimen is a part of right maxilla of *Crocodylus palaeindicus* with alveoli for the attachment of 6 maxillary teeth and 5 premaxillary teeth. The alveoli indicate that the teeth were circular in shape, the 4th premaxillary tooth was the biggest in diameter and in the series and the rest were relatively smaller. The mid line on the palatal view is prominent. Dorsally the specimen is characterized by deep pitting; the premaxillary portion is more elevated in the side view. Upper cranial profile is slightly convex. The specimen indicates that the skull of the species was short and broad.

TAPHONOMIC ATTRIBUTES

Crocodile is one of the large predators with sharp teeth and strong jaws. In spite of enormous water bodies in the Narmadā Valley, stagnant and actively flowing, its fossils are scanty; parts of skulls, jaws, vertebrae and isolated teeth are generally found in small quantities. The specimen is

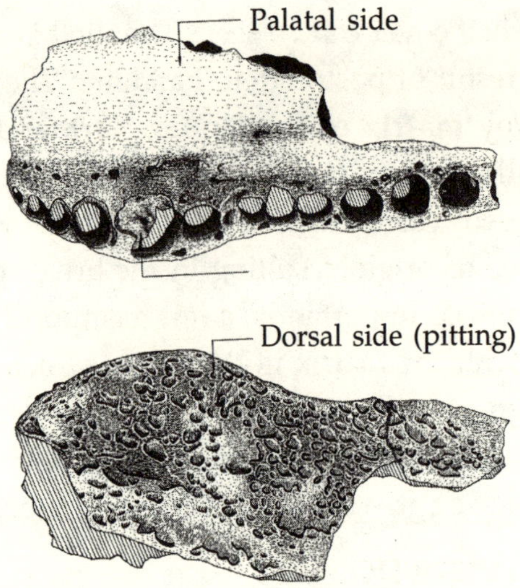

fig. 6.2 NMD/1239/DC Part of maxilla of *Crocodylus palaeindicus*

stained red because of the iron oxide minerals in the soil from which it has been collected. The pattern of breakages indicates that the specimen has been affected by fluvial transport and drifting.

Some concretionary materials are attached on the palatal side of the skull.

Specimen no. NMD/13/DC

Locality	:	Devakachār area
Horizon	:	Sandy pebbly gravel
Identification	:	Lower mandible with molar

SYSTEMATIC PALAEONTOLOGY

Phylum	:	Vertebrata
Class	:	Mammalia Linnaeus, 1758
Order	:	Proboscidea Illiger, 1811
Family	:	Elephantidae Gray, 1821
Genus	:	*Elephas* Linn., 1758,
Species	:	*Elephas hysudricus* Falconer and Cautley, 1845

DESCRIPTION (*fig.* 6.3)

The specimen is a right lower molar. Seven rows of cones are well preserved with three anterior and one posterior cone broken and missing. The first preserved cone from the anterior side has crenulated thick enamel and the second one is mildly crenulated. The posterior cones are not crenulated and are less worn. The valleys are narrow and deep.

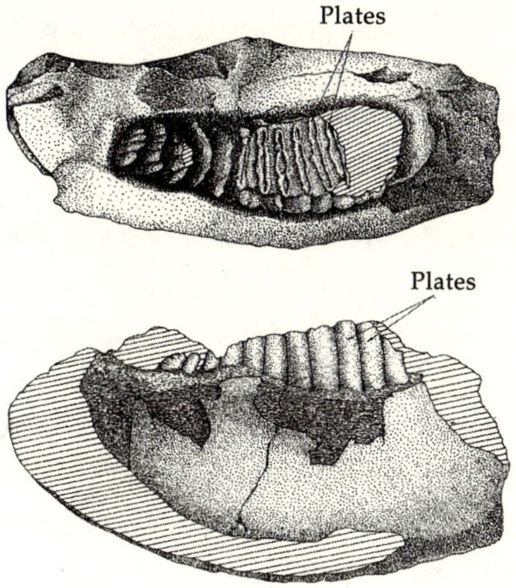

fig. 6.3 NMD/13/DC Lower molar of *Elephas hysudricus*

MEASUREMENTS

Maximum length of the mandible	:	315.8 mm
Maximum breadth of the mandible	:	151.2 mm
Length of the occlusal surface	:	225.7 mm
Breadth of the occlusal surface	:	90.0 mm
Crown height	:	9.5 mm

TAPHONOMIC ATTRIBUTES

The specimen exhibits several mechanical breaks that must have occurred due to excavation and its subsequent handling as is evident by the freshness of the breaks. Longitudinal breaks along the weak zones of the plates throughout the length of the molar can be seen. The breaks on the anterior plates are shallow but are deeper on the posterior ones. Evidence of flaking is also present on three of the plates though not vey prominent.

Specimen no. NMD/11/DC

Locality	:	Barmanghāṭ
Horizon	:	Sandy pebbly gravel
Identification	:	Molar

SYSTEMATIC PALAEONTOLOGY

Phylum	:	Vertebrata
Class	:	Mammalia Linnaeus, 1758
Order	:	Proboscidea Illiger, 1811
Family	:	Elephantidae Gray, 1821

Genus	:	*Stegodon* Falconer and Cautley, 1847
Species	:	*Stegodon insignis ganesa* (Falconer and Cautley), 1845

DESCRIPTION (*fig.* 6.4)

The specimen, a right molar, is represented by 8 lophs (plates) which are cumulatively broader at the base and converge to become narrower towards the occlusal end. The specimen has an elliptical shape and the lophs are securely embedded in the jaw. The plates are gradually bending towards the base of the preserved end. The crown is well preserved. The occlusal surface has got 2 rows of cones separated by a narrow valley. The first row of cones bears 5 conelets whose enamel is not crenulated. The 2nd row of cones bears 6 conelets.

MEASUREMENTS

Maximum length	:	275.4 mm
Maximum breadth	:	170.0 mm
Maximum thichness	:	95.3 mm

TAPHONOMIC ATTRIBUTES

The specimen is well preserved. Several cracks across the enamel of the plate indicate long exposure to sub-aerial weathering and also probably thermal fractures. The lingual side below the cones has a number of pit marks perhaps made naturally. The base of the jaw has some shallow

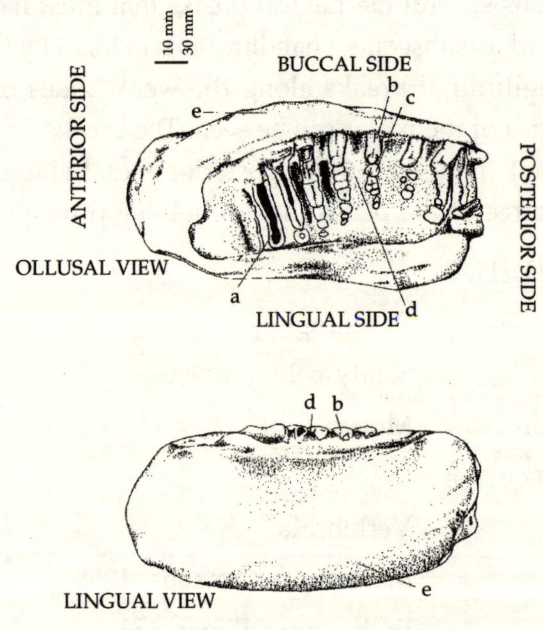

fig. 6.4: NMD/11/DC Molar of *Stegodon insignis*

cracks in the alignment of the bone. The species comes from sandy peddly gravel and was collected from the top of a dyke scattered with Middle Palaeolithic flakes and boulders, indicating that it must have been transported along the river during high energy environment. When the water receded, it got deposited at the top of the dyke, a case of a lag deposit.

The buccal side has slight indication of a lateral stress in the bone resulting in some shallow depressions and ridges. The specimen while in transport must have come in contact with big boulders but because of its size and weight it withstood the impact. Furthermore, the transportation of a bone, so heavy and big, suggests a high energy environment that must have prevailed in the area.

Specimen no. NMD/1051/DC

Locality	:	Guarighāṭ
Horizon	:	Sandy pebbly gravel
Identification	:	Lower dentition with mandibular symphysis

SYSTEMATIC PALAEONTOLOGY

Phylum	:	Vertebrata
Class	:	Mammalia Linnaeus, 1758
Order	:	Artiodactyla Owen, 1847
Family	:	Hippopotamidae Gray, 1821
Genus	:	*Hexaprotodon* Falconer and Cautley, 1836
Species	:	*H. palaeindicus* Falconer and Cautley, 1836

DESCRIPTION (*fig.* 6.5)

The specimen is represented by a partially preserved mandible of *Hexaprotodon palaeindicus* (the six incisored hippopotamus). The mandible bears six hollow sockets for the incisors, two broken canines, four pre-molars and three molars on the left horizontal ramus and the remnants of four pre-molars on the right ramus (p_1-p_4).

The diagnostic characters of this particular specimen are the elongated mandibular symphysis — a habitat character for holding the six incisors. The lateral extension of the symphysis is designed to hold the large canines which are oval in cross-section. The great height of the horizontal ramus and the convex shape of the lateral side of the ramus starts from the fourth pre-molar and sharply bends laterally outwards in order to accommodate the canines. The base of the mandibular symphysis is slightly convex with

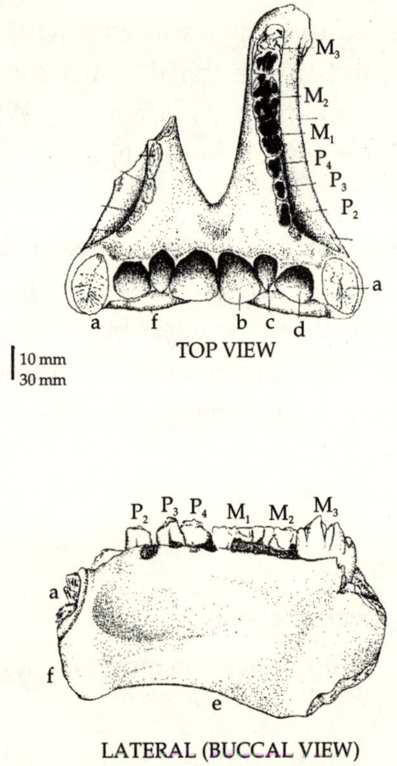

a. canine (broken)
b. incisor
c. incisor
d. incisor
e. Horizontal ramus
f. Mandibular symphysis

fig. 6.5: NMD/1051/DC Mandible (with symphysis) of *Hexaprotodon palaeindicus*

a dent in the middle. The overall appearance of the mandible is rather massive and robust. The interior part of the alveolar arch is rather deep and V-shaped with steep sides which turn outwards at the base of the jaw, providing more space in its interior part. The dental character shows the evidence for the ruminating behaviour of the animal. The pre-molars are single cusped and the molars are double cusped with flower patterns. The teeth starting from P2 to M3 show a gradual increase in the occlusal surface area in sharp contrast to those of other ruminants such as bovids and equids, where the occlusal surface area decreases from the pre-molars to the last molars. The crown height is high. The M3 appears to have been in growth process at the time of death, indicating that the animal was not very old.

The number of incisors of the specimen, and those of all others related to this genus, has attracted much debate on its taxonomy from earlier researchers like Falconer and Cautley (1836) who on the basis of the number of incisors (six and four) grouped them into two sub-genera of the genus *Hippopotamus*. Owen (1845) found it more appropriate to group the six incisored ones under the genus *Hexaprotodon* and the four incisored ones under the genus *Hippopotamus.* Lydekker (1882), Hooijer (1950) and Biswas and Dassarma (1984) are of the opinion that the classification based upon the number of incisors could differentiate only at the specific level and therefore they put the two mentioned genera together under *Hippopotamus* alone. Colbert (1935), on the basis of the number of incisors and cranial features, concluded that they represented two separate genera, viz. *Hexaprotodon* and *Tetraprotodon* respectively. This view of Colbert was followed by Deraniyagala (1956), Badam and Kajale (1977 a, b) and Nanda (1978).

Some scholars are of the view that the decrease in the number of incisors has got something to do with the evolutionary stages of the *Hippopotamus.* In the specimen studied here, the second incisor socket shows a much reduced diameter and depth, obviously depicting that the size of the incisors was much smaller than that of other four. Moreover, these sockets appear as if they have been pushed towards the interior margin of the alveolar process. The smallness of the second incisors and the process of reduction finally reached a stage when the incisors got diminished and disappeared giving rise to a new species with four instead of six incisors. This was later termed as *Tetraprotodon* as opposed to the *Hexaprotodon* with six incisors. It may be noted here that the intermediate stage between the six and four incisored hippo has also been reported which has been termed as *Pentaprotodon* — a new genus filling up the evolutionary gap as far as the number of incisors are concerned. This new discovery was reported as an abnormal *Hexaprotodon sivalensis* with the name *Pentaprotodon sivalensis* from Pinjor Formation of the Upper Śiwāliks of north-west India (Sahni and Khan, 1961). This view was supported by Biswas and Dassarma (1984) Badam and Kajale (1977 a, b).

From the discoveries made in India of the different genera of the specimen, a considerable amount of evolutionary significance can be drawn. Lydekker (1882), Hooijer (1950), Badam and Kajale (1977 a, b) and Biswas and Dassarma (1984) are of the opinion that *Hexaprotodon palaeindicus* of the Narmadā Pleistocene probably evolved from *Hexaprotodon sivalensis* through an intermediate genera *Hexaprotodon namadicus.*

MEASUREMENTS

Maximum length	:	305.3 mm
Linear distance between the lateral side of the two canines	:	295.0 mm

Maximum thickness of the canine	:	55.0 mm
Crown of P2	:	Height : 20.0 mm; Length : 22.4 mm
Crown of P3	:	Height : 22.1 mm; Length : 22.3 mm
Crown of P4	:	Height : 22.7 mm; Length : 24.1 mm
Crown of M1	:	Height : 22.8 mm; Length : 37.3 mm
Crown of M2	:	Height : 23.0 mm; Length : 39.0 mm
Crown of M3	:	Height : 39.1 mm;Length : 48.6 mm
Height of the ramus at M3	:	150.0 mm
Height of the ramus at P3	:	130.1 mm
Height of the ramus at the canine	:	122.7 mm

TAPHONOMIC ATTRIBUTES

The specimen is well preserved and heavy with almost complete left alveolar region. Sixty per cent of the two canines are well preserved. The damage that has occurred is due to post-excavation handling of the fossil. At the lower end of the mandibular symphysis there is red patination covering ¾th of the mandible indicating the presence of iron oxide in the soil. On the left buccal side there are several cracks following the major orientation of the bone and one shallow vertical crack below the line of the 3rd pre-molar, all these have contributed to the flaking of some bone material at the lower end. On the dorsal side of the 2nd incisor on either side, shallow cracks have appeared which have not, however, caused much damage to the specimen. The hind part of right jaw is broken off and has a number of cracks similar to weathering stage II or even stage III. However, some cracks appear to have been caused due to handling.

Sandy pebbly gravel matrix is still attached to the mandible and teeth but indications on the state of preservation are that the specimen has not undergone much rolling. It, therefore, can be treated as being autochthonous to the local channels.

Specimen no. NMD/713/DC

Locality	:	Devakachār
Horizon	:	Sandy pebbly gravel
Identification	:	Left humerus of *Hexaprotodon namadicus*

SYSTEMATIC PALAEONTOLOGY

Phylum	:	Vertebrata
Class	:	Mammalia Linnaeus, 1758
Order	:	Perissodactyla Owen, 1847
Family	:	Hippopotamidae Gray, 1821
Genus	:	*Hexaprotodon* Falconer and Cautley, 1836
Species	:	*Hexaprotodon namadicus*

DESCRIPTION (*fig.* 6.6)

The specimen is massive and complete. The proximal end of the fossil is broader than the distal end. The head is protruding medially and has a short neck. The lateral tuberosity rises about 50 mm above the groove which is filled up by soil matrix. The medial tuberosity is also well preserved. Relatively, the shaft is short and stout with its circumference decreasing towards the base of the broader distal end. It has a prominent deltoid tubercle on the posterior lateral side, about 1/3rd of the distance from the proximal end to the distal end. The olecranon fossa is deep, filled up with matrix and appears to be triangular in outline. The lateral and medial condyles are well rounded with a prominent intercondylar groove. The lateral condyle is larger than the medial condyle. The coronoid fossa above the condyles is shallow and filled up with matrix.

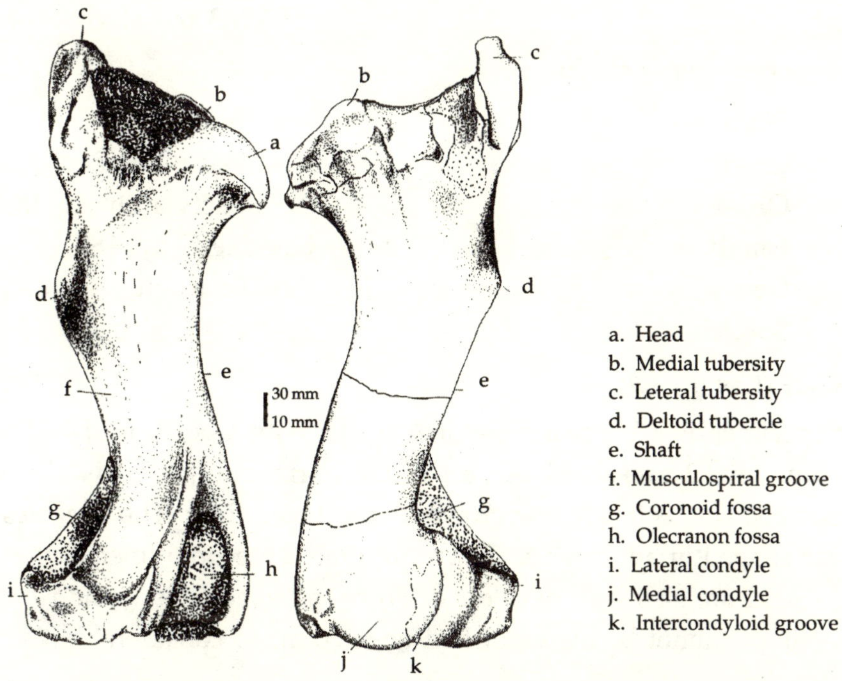

fig. 6.6: NMD/713/DC Left humerus of *Hexaprotodon namadicus*

MEASUREMENTS

Maximum length of the specimen	:	392.8 mm
Horizontal distance from the lateral tubercle to the lateral tip of the head	:	144.0 mm
Breadth of the shaft at the deltoid tubercle	:	94.2 mm
Breadth of the condyles	:	92.2 mm

TAPHONOMIC ATTRIBUTES

The specimen has been weathered considerably and has lost its colour. Possibly the specimen got exposed to subaerial weathering process for a long time. The matrix attached is hard gravelly material indicating that it must have been a part of lag deposit in the upper regimes of sedimentation. Several shallow cracks in the middle of the trochlea and below the head and also on the shaft indicate some stress like water logging, trampling or sediment load that this specimen must have undergone. Evidence of rodent gnawing and bioturbation is seen at the proximal side near the greater tuberosity and below it. Two transverse fractures at the proximal half of the specimen are the handiwork of mechanical breakage.

Specimen no. NMD/548/DC

Locality	:	Talayyāghāṭ
Horizon	:	Sandy pebbly gravel
Identification	:	Left mandible with P4, M1, M2, M3 and alveolus of P3.

SYSTEMATIC PALAEONTOLOGY

Phylum	:	Vertebrata
Class	:	Mammalia Linnaeus, 1758
Order	:	Artiodactyla Owen, 1847
Family	:	Bovidae Gray, 1821
Genus	:	*Axis* Erxleben
Species	:	*Axis axis* Erxleben

DESCRIPTION (*fig.* 6.7)

The specimen is a left mandible of *Axis axis* with teeth from P4-M3 and alveolus for P3 preserved. A part of the horizontal ramus is also well preserved. The teeth are low crowned and heterodont. The fossettes are saucer shaped in appearance. The protoconid, entoconid, metaconid, and hypoconid are also well developed. From the buccal view, the ribs and the valleys cannot be distinguished on account of considerable wear of the specimen.

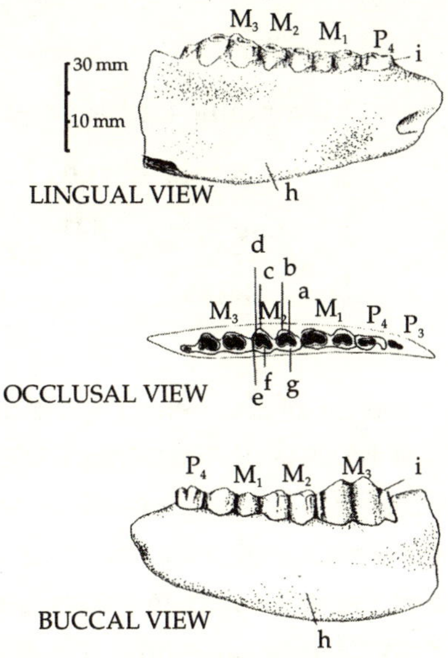

a. Pre-fossette
b. Metaconid
c. Entocoind
d. Post-fossette
e. Hypoconulid
f. Hypoconid
g. Protocoind
h. Horizontal ramus
i. Crown

fig. 6.7: NMD/548/DC Left mandible of *Axis axis*

MEASUREMENTS

Maximum length of ramus	:	88.8 mm
Height of ramus at M3	:	37.6 mm
Height of ramus at P3	:	25.8 mm
Maximum thickness of the mandible	:	10.3 mm
Length of P3	:	9.1 mm
Length of M1	:	15.6 mm
Length of M2	:	17.2 mm
Length of M3	:	21.9 mm

TAPHONOMIC ATTRIBUTES

The specimen is well preserved with teeth P4-M3 and alveolus for P3. Though the specimen has been rolled considerably, there are only two cracks in the specimen, which are narrow but deep, the vertical one below the first cusp of M2 and the other one below the first cusp of M1 right up to the base of the jaw. On the lingual side there are indications of long striation marks. Decolouration has occurred on both sides of the jaw.

Specimen no. NMD/597/DC

Locality	:	Talayyāghāṭ
Horizon	:	Sandy pebbly gravel
Identification	:	Thoracic vertebra

SYSTEMATIC PALAEONTOLOGY

Phylum	:	Vertebrata
Class	:	Mammalia Linnaeus, 1758
Order	:	Artiodactyla Owen, 1847
Family	:	Bovidae Gray, 1821
Genus	:	*Bos / Bubalus*

DESCRIPTION (*fig.* 6.8)

The vertebra is complete except for the tip of the neural spine and the left transverse process. The spine tends to be long and slanting backwards and it appears that only 1/3rd of its total length is preserved. The neural canal is completely blocked by cementing material and is somewhat oval in cross-section. The body of the centrum is big and oval in cross- section. The transverse process is short and curved upwards and the anterior and posterior coastal articular facets are well preserved. Pre- and post-zygapophyses are well preserved, with the former being more or less triangular and short while the latter, elongated. There are three prominent ridges on the body of the centrum joining the cranial and caudal articular facets.

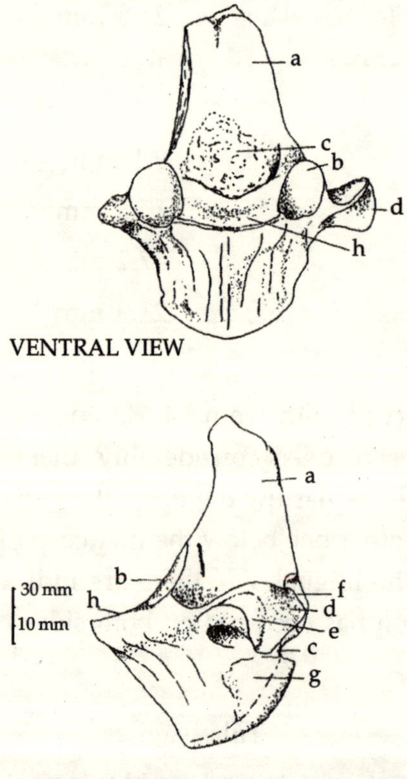

a. Neural spine
b. Post zygapophysis
c. Neural canal
d. Transverse process
e. Coastal articular facet
f. Pre zygapophysis
g. Facet for rib tubercle
h. Centrum

fig. 6.8: NMD/597/DC Thoracic vertebra of *Bos/Bubalus*

MEASUREMENTS

Base of the centrum to the tip of the broken neural spine	:	148.0 mm
Maximum breadth (from one lateral tip of the transverse process to another)	:	130.6 mm
Length of the cranial articular facet	:	62.8 mm
Breadth of the cranial articular facet	:	53.9 mm
Length of the cranial side of neural spine	:	22.2 mm
Breadth of the cranial side of the neural spine	:	33.4 mm
Length of the caudal articular facet	:	57.3 mm
Breadth of the caudal articular facet	:	62.4 mm

TAPHONOMIC ATTRIBUTES

The specimen is covered by hard gravelly matrix. The proximal part of the neural spine is broken and so is the left transverse process. The breakage appears fresh. The portion below the right transverse process and the lateral side of the centrum is flaked off by mechanical means. There are no evidences of bioturbation, gnawing or even cracks on the fossil. The specimen might have been a part of the lag deposit after being rolled down from the place of origin.

Specimen no. NMD/775/DC

Locality	:	Guarighāṭ II
Horizon	:	Sandy pebbly gravel
Identification	:	Right maxilla with M1, M2 and alveolus for P4

SYSTEMATIC PALAEONTOLOGY

Phylum	:	Vertebrata
Class	:	Mammalia Linnaeus, 1758
Order	:	Artiodactyla Owen, 1847
Family	:	Bovidae Gray, 1821
Genus	:	*Bos / Bubalus*

DESCRIPTION (*fig.* 6.9)

The specimen is a right juvenile maxilla of a bovid with M1 and M2 preseved fully and the alveolus for P4 present. Part of the palatine can also be seen. The basal pillar has reached full height. The protoconid, entoconid, metaconid and hypoconid are also distinctly visible and so are the fossettes. The P4 seems to have already erupted though it is broken

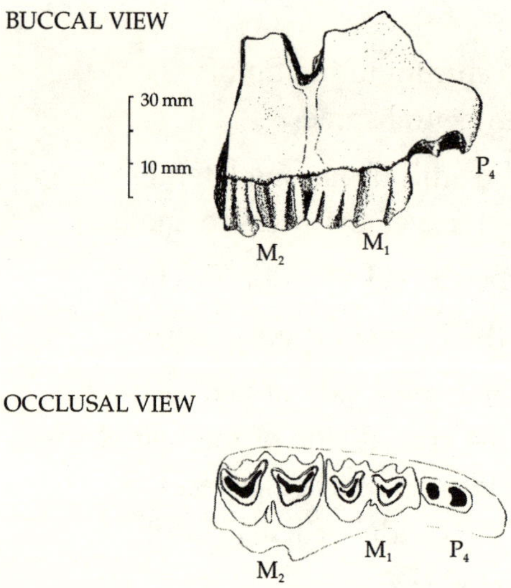

fig. 6.9: NMD/775/DC Right maxilla of *Bos/Bubalus* (juvenile)

off. Each of the molars, from the buccal view, show 5 ribs and 4 valleys. The dentition is heterodont and gives a step-like appearance. The specimen is associated with a lag deposit which might explain its excellent preservation.

MEASUREMENTS

Maximum height of the mandible	: 74.1 mm
Length of M1	: 27.8 mm
Height of M1	: 16.2 mm
Length of M2	: 30.3 mm
Height of M2	: 15.1 mm
Length of alveolus for P3	: 15.2 mm
Breadth of alveolus for P3	: 7.6 mm

TAPHONOMIC ATTRIBUTES

Though the specimen is well preserved. It may have belonged to a juvenile individual. It was found in pebbly cobbly matrix and gives indication of having been re-deposited. However, the cusp pattern is not damaged and worn out. A major part of the palate is broken, probably by rolling and transporation.

Specimen no. NMD/808/DC

Locality	:	Belghāṭ
Horizon	:	Sandy pebbly gravel
Identification	:	Partial skull

SYSTEMATIC PALAEONTOLOGY

Phylum	:	Vertebrata
Class	:	Mammalia Linnaeus, 1758
Order	:	Artiodactyla Owen, 1847
Family	:	Bovidae Gray, 1821
Genus	:	*Bibos* Smith
Species	:	*Bibos gaurus* Smith

DESCRIPTION (*fig.* 6.10)

The specimen is represented by posterior part of the frontal and the base of the skull, a considerable portion of the occiput, temporal grooves and the cavity for the right horn core. Most of the features on the occipital region are well preserved including the occipital condyles, foramen magnum and nuchal crest with a very prominent superior nuchal line. The foramen magnum which is roughly round in shape and broad is filled up by sandy matrix. The occipital condyles are triangular and located at the posterior end of the skull, protruding outwards. The mastoid processes are half broken and only their bases are preserved. The parieto-temporal, parieto-occipital and fronto-parietal sutures are well marked. A major portion of the frontal bone is missing while only the posterior part of its half is intact. Most of the frontal bones are intact with a part of the

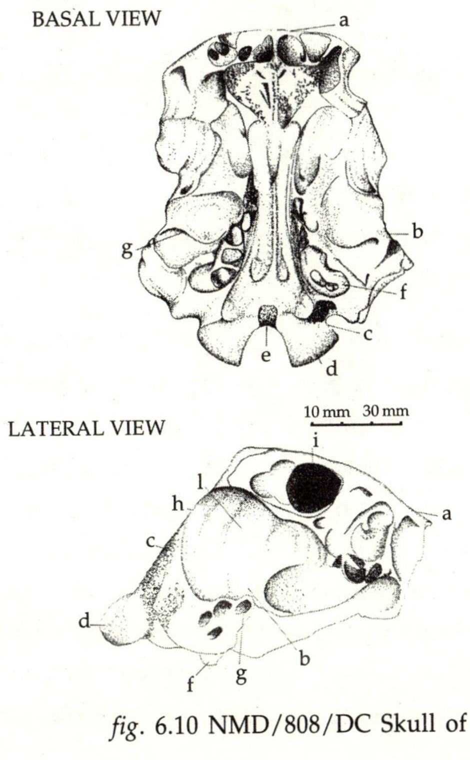

a. Frontal bone
b. Temporal bone
c. Occipital bone
d. Occipital condyle
e. Foramen magnum
f. Mastoid process
g. Tympanic bulla
h. Nuchal crest
i. Temporal groove
j. Cavity for the root of the horn core.

fig. 6.10 NMD/808/DC Skull of *Bibos gaurus*

tympanic cavity. The temporal fossae are deep. The zygomatic extension of the temporal bone is broken at its base.

The preserved portion of the skull is massive and huge with the skull bones being considerably thick. The base of the skull bears a number of fossae and cavities for blood vessels, nerves and tissues. The entire skull has become dark grey in colour indicating its association with a manganiferous soil environment. This specimen has been earlier described by Badam and Grigson (1990).

The features of the specimen show certain similarities with those of *Bos namadicus.* Pilgrim (1939) in his monograph on the *"Fossil Bovidae of India"* has pointed out certain characteristics in which *Bos* differs from *Bibos.* In *Bibos* the summit of the occipital crest lies far below the lower border of the horncores and above it the parietals and supraoccipitals occupy a relatively wide area. The temporal fossae open widely into that portion of the occiput which extends far above the occipital crest and includes both the supraoccipitals and the parietals; also the posterior extremities approach one another more closely. The longitudinal extension of the frontals and the part of the face anterior to them are approximately equal.

MEASUREMENTS

Maximum length from occipital condyles to the anterior part of frontal region	: 322.7 mm
Maximum breadth of the skull at the temporal region	: 261.3 m
Distance between the two condyles	: 50.8 mm
Length of the mastoid process	: 60.2 mm
Distance between the two mastoid processes	: 158.9 mm

TAPHONOMIC ATTRIBUTES

It is a rare specimen as only half a dozen skulls of this species have been found in the valleys of Narmadā and Son. The foramen magnum, parietal part and some of the basi-occipital parts have hard sandy pebbly matrix attached. The edges of the skull (parietal, frontal, supraoccipital regions) are damaged. Even though the specimen is well preserved, it is obvious that is has undergone some transportation that has resulted in its breakage of ex-occipital region and some parts of the frontal region. Some of the portions of the connective tissues of the occipital condyles are visible but this may be due to erosion or abrasion. On the right side of the basi-occipital region there are small hollows recalling the action of bioturbation but one cannot be sure. The specimen definitely comes from a lag deposit and is in a somewhat good state of preservation.

Specimen no. NMD/1322/DC

Locality	:	Gurāriāghāṭ
Horizon	:	Sandy pebbly gravel
Identification	:	Lower molar

SYSTEMATIC PALAEONTOLOGY

Phylum	:	Vertebrata
Class	:	Mammalia Linnaeus, 1758
Order	:	Perissodactyla Owen, 1847
Family	:	Equidae Gray, 1821
Genus	:	*Equus* Linn., 1758
Species	:	*E. namadicus* Falconer and Cautley, 1849.

DESCRIPTION (*fig.* 6.11)

The specimen is a lower molar of *Equus namadicus*. The occlusal part is weathered and fairly worn out and only traces of metastylid can be seen. From the buccal view, the two ribs and valley can be prominently distinguished. Dentition is hypsodont with a slight curve in the crown length. The specimen has turned black in colour indicating the presence of manganese in the soil.

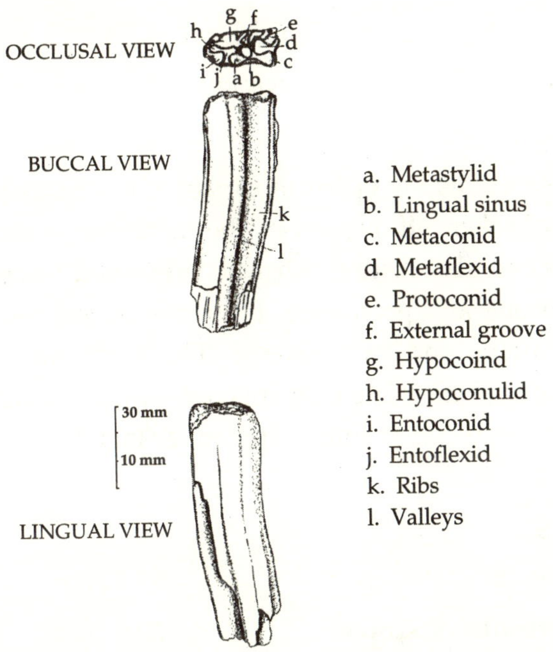

fig. 6.11: NMD/1322/DC Lower molar of *Equus namadicus*

MEASUREMENTS

Maximum length of the molar	:	82.1 mm
Maximum breadth of the molar	:	35.3 mm
Maximum thickness of the molar	:	22.3 mm

TAPHONOMIC ATTRIBUTES

The specimen is the right lower molar of *Equus namadicus.* The crown is damaged but the enamel structure can still be made out. Parts of the ridges of enamel of the crown are longitudinally broken at different portions in the inner side. There is also a shallow crack at its side depicting some kind of thermal fracture. The root is broken and bears evidence of weathering stage I.

Specimen no. NMD/498/DC

Locality	:	Devakachār
Horizon	:	Sandy pebbly gravel
Identification	:	Upper molars

SYSTEMATIC PALAEONTOLOGY

Phylum	:	Vertebrata
Class	:	Mammalia Linnaeus, 1758
Order	:	Perissodactyla Owen, 1847
Family	:	Equidae Gray, 1821
Genus	:	*Equus* Linn., 1758
Species	:	*E. namadicus* Falconer and Cautley, 1849

DESCRIPTION (*fig.* 6.12)

The specimen represents the upper molars of *Equus namadicus.* One of the molars is partly broken off. However, the metastyle is prominent. On a closer look, one can also distinguish well preserved fossettes, entocone, metacone, hypocone and protocone. From the buccal view, each of the molars has two ribs and two valleys. The teeth are high crowned (hypsodont). The specimen is partly covered with cementing matrix. The colour has become reddish indicating its association in an iron rich environment.

MEASUREMENTS

Maximum length of crown	:	87.2 mm
Maximum breadth of crown	:	20.9 mm
Maximum thickness of crown	:	14.3 mm

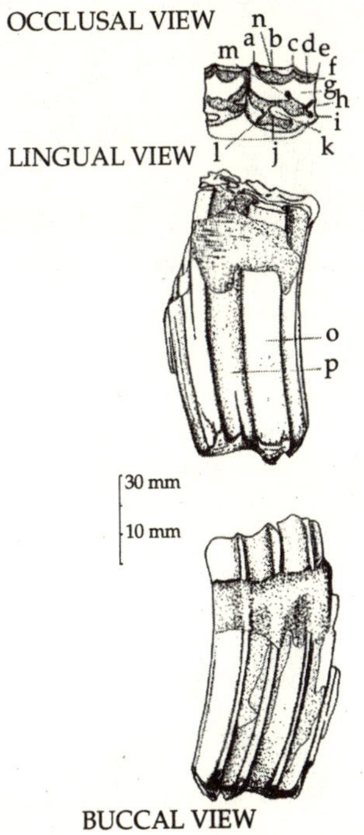

a. Parastyle
b. Anterior interstylar face
c. Mesostyle
d. Posterior interstylar face
e. Metastyle
f. Metacone
g. Post-fossette
h. Metaconule
i. Hypocone
j. Protocone
k. Plicaballin fold
l. Paraconule
m. Prefossette
n. Paracone
o. Rib
p. Valley

fig. 6.12: NMD/498/DC Upper molars of *Equus namadicus*

TAPHONOMIC ATTRIBUTES

The specimens are the left upper molars of *Equus namadicus.* Much of the bone material on the buccal side has been eroded exposing the ridge and valley structure. Crown surface is reddish in colour and there are several horizontal cracks on the other side of the molars which have damaged the teeth. The texture of the teeth indicates that they might have remained exposed for a long time. The wear on the occlusal surface indicates an old age for the specimens.

7

Taphonomy and Palaeoecology

Taphonomy

BEFORE discussing the topic in detail with respect to the Narmadā Valley, let us add a brief introduction to taphonomy, a relatively new branch of palaeontology, with relevant examples which will help us understand better the mode of deposition of the Narmadā Alluvium.

The term "Taphonomy" was proposed by a Russian palaeontologist, J.A. Efremov in 1940 for the study of the transition of animals from the stage of biosphere to that of lithosphere (Efremov, 1940). The word has a Greek origin, *taphos* means burial and *nomos* means law. Literally therefore, taphonomy means "laws of burial." It is a process in which animals pass out of different parts of the biosphere, get fossilized and become part of the lithosphere. In fact, during the last few years the application of taphonomy and palaeoecology has so widened that both are now considered as archaeology's sister disciplines (Gifford, 1981).

It may be mentioned that taphonomy established itself in palaeontology primarily as a sub-discipline of palaeoecology but it has evolved into a much broader study of the ways in which preservation affects the fossil record (Behrensmeyer and Kidwell, 1985). In the US, where such studies have received tremendous attention, there has been a change in emphasis in recent years from descriptive taphonomic studies of fossil assemblages to more experimental, process-oriented multidisciplinary investigations (Behrensmeyer, 1990). This research is largely helping to resolve problems in evolution, biogeography, biostratigraphy and palaeo-environmental reconstruction, and thus to a better overall understanding of the fossil record. However, continuing research in modern analogues and theoretical modelling is critical to the further understanding of taphonomy as is the interaction between biological and geological processes.

Taphonomy can be divided into two fields, biostratinomy and diagenesis. Biostratinomy includes the circumstances occurring between the death of an individual and its subsequent burial whereas diagenesis includes the effects upon the interred remains brought about by burial conditions and subsequent exposure until the time of discovery. Gilbert (1979) correlates these fields with events in fossil history as shown below.

	Palaeontological sub-disciplines	*Events in fossil history*
	Diagenetic studies	Discovery
Taphonomy	Biostratinomy	Final Burial
	(Palaeoecology)	Death
		Birth

The various stages in taphonomical processes from the death of an animal to its recovery as a fossil have been diagrammatically shown as below:

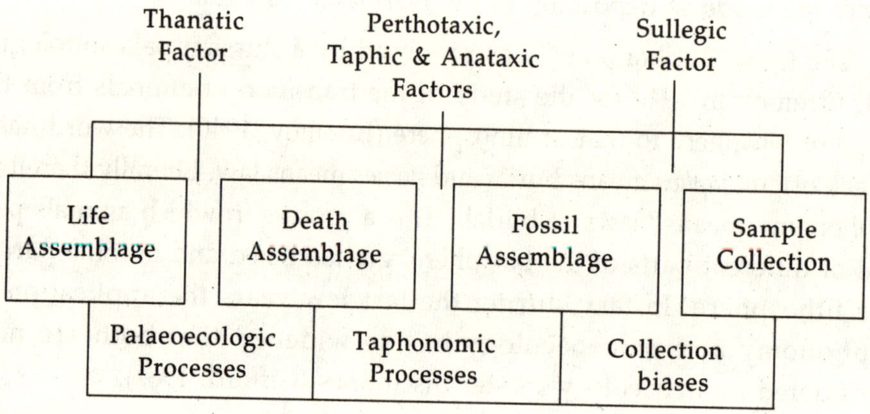

Diagrammatic History of Taphonomy

Explanations for the various terms used in the above diagrammatic history are given below:

Life Assemblage	Biocoenosis	=	population of living animals
Death Assemblage	Thanatocoenosis	=	population of dead animals
Fossil Assemblage	Perthotaxic	=	variables affecting bone preservation subsequent to death but prior to burial
	Taphic	=	mode of burial of animal remains
	Anataxic	=	site erosion, nature and rate of sedimentation, depth of burial, sediment chemistry, bioturbation, etc.
Collection.............................	Sullegic factor	=	manner of collecting fossils and the collecting bias.

The sum total of the fossil forming processes involved during the passage from the biosphere into the lithosphere, occur as a result of many interrelated geological, biological and chemical phenomena. The different events in such a phenomena are generally considered to provide insights into the causes of death, decomposition and disarticulation, weathering and transport, diagenetic factors, collection of samples and documentation of the animals.

Some of the problems generally connected with fossilization and the role of taphonomy in solving them are given below:

1. Bones from older strata and those from younger strata react differently to the process of fossilization. In the former the state of preservation of fossils is often poor and the number of individuals fewer because chances of destruction increase with time. However, it is the fauna from the younger strata that are more often subject to destruction on account of tectonic events, floods, ploughing, etc.
2. Faunal assemblages, especially those of vertebrates, may not always be autochthonous. Although a complex of terrestrial mammals found together in one place is usually called a fauna (a term implying that the animals were associated in life), such a complex may have come together accidentally either at or after death.
3. The sudden appearance in the fossil record of a new fauna, without its ancestors in the preceding strata, may or may not reflect a genuinely new group of species. This may be an artefact of preservation. Sometimes strata may be subjected to disturbances resulting in the lack of a continuous record of fossil species.
4. Various parts of skeletons are preserved and fossilized differently in different sedimentary environments. Sometimes the assemblages of a single synchronic and sympatric community may be mistaken for different fauna because of different sedimentary processes and different methods of sampling.
5. In case bones of different animals are found together in a particular sedimentary deposit it is difficult to say whether these bones were buried together or transported together or whether the animals lived and died at the same site.

The problems outlined above have a direct or indirect bearing on palaeoecology and the sedimentary environment and can be understood better by studying taphonomy which is generally regarded as a prerequisite to palaeoecology. Apart from giving precise information on palaeoecology, the study of taphonomy also helps to solve the problems of mode of occurrence, taxonomic diversity and paucity of small mammals (if any) and explains various aspects regarding the concentration of fossils, patterns of disarticulation, varieties of damage, etc.

In recent years, taphonomy has been increasingly used in archaeology and geology for a more precise reconstruction of chronology and the palaeo-environment. Since taphonomy throws light on the autochthonous/allochthonous nature of fossils one can know whether these are/are not related to the sediments yielding them. However, in the Indian context discrepancies have crept in while relating fossils to sediments. For example, in the Śiwāliks of NW India, where earlier collections came from mixed contexts, without a clear-cut stratigraphic control (Badam and Rajaguru, 1991) the chronological framework established was somewhat erroneous till the methods of physical and chemical dating became available. In several cases the matrix enclosing the fossils is different from the lithological units from which the fossils are extracted (Saxena, 1968). A detailed taphonomy of the Śiwālik fossils has now thrown new light on the provenance and mode of deposition of these fossils which in turn have an important bearing on the biochronology of the Śiwālik sediments.

In some archaeological sites the dating of artefacts is also problematic. Recent taphonomic work involving density and flow studies of geological sediments in the rubble deposits at the Middle Palaeolithic site of Samnapur in Madhya Pradesh, has shown that the artefact bearing colluvial rubble is a product of debris flow processes and artefacts occur in a primary context only in the upper part (30 cm) of the rubble (Ahsan, 1993). In the Ghoḍ Valley near Ināmgāon, however, a clear-cut distinction has been observed in the gravels yielding the Middle Palaeolithic scrapers and flakes and the overlying gravels yielding Upper Palaeolithic blades (Badam, 1985), but such instances are rare.

A similar difficulty arises in respect of fossils. Our extensive work in the Central Narmadā Valley has shown that taphonomically two modes of deposition can be detected in the Devakachār area of Narsinghpur Dist., Madhya Pradesh. The majority of the fossils found in the lag deposit of the bouldery and cobbly horizon are fragmented and seem to have undergone some transportation. Fresh and unabraided fossils are derived from the overlying silty sandy horizon (channel fill and overbank alluvial fill deposits) and probably the animals represented by these fossils were autochthonous.

Pl. 7.1: Taphonomy Experiments
in the Central Narmadā Valley (Stage I).

Pl. 7.2: Taphonomy Experiments
in the Central Narmadā Valley (Stage II).

Pl. 7.3: Taphonomy Experiments
in the Central Narmadā Valley (Stage III).

Pl. 7.4: Taphonomy Experiments
in the Central Narmadā Valley (Stage IV).

An interesting case of a re-deposited fossil which raises some doubts about its *in situ* position, is that of the partial cranium of *Homo erectus narmadiensis* found in a channel bar environment at Hathnorā in Madhya Pradesh (Sonakia, 1984). The cranium is heavily mineralized and has been found associated with Late Acheulian artefacts. Its position in the bar deposits combined with taphonomic features indicate that the cranium was re-deposited in this section after having been transported over some distance (Badam, 1989). This aspect if overlooked would give erroneous results regarding the dating of the cranium. Thus the sedimentary context and the associated cultural materials warrant a detailed taphonomic study to arrive at a precise chronology and palaeo-environmental history of the site.

It may be stressed here that research in the new discipline of taphonomy and its related branches on the Quaternary fossils of India, was initiated by the author at the Deccan College after a preliminary study of a small collection of fossils from the Central Narmadā Valley (Badam et al., 1986). Subsequently, additional aspects of this fascinating study were undertaken on collections from Karīmnagar, Andhra Pradesh (Badam and Jain, 1988), Son valley in Uttar Pradesh (Badam et al., 1989) and other areas in Central and Western India (Badam, 1990, 1996).

Studies on the decomposition and scatter patterns of modern animals have been carried out at various National Parks in Africa. They provide a modern analogue for reconstructing past taphonomic stages leading to their preservation and subsequent fossilization of the skeletal remains. Studies on similar lines have been carried out at the Mudumalai Sanctuary in Tamil Nadu and the Narmadā Valley by the author (*Pl.* 7.1 to *Pl.* 7.4). This preliminary effort is rather broad based but a rigorous follow-up study will help throw more precise light on the chronology and depositional environment of the fossils in relation to archaeological, geological and palaeontological sites in India.

The Narmadā Ecology

Ecologically, the faunal biota represented by the fossils can be summarized in the table given below:

Table 6: Fossil Communities in the Central Narmadā Valley

Primary consumers	*Hexaprotodon palaeindicus* *Hexaprotodon namadicus*	*Equus namadicus, Stegodon insignis ganesa, Elephas hysudricus, Bos namadicus, Bubalus palaeindicus, Bibos gaurus, Boselaphus tragocamelus, Antilope cervicapa, Cervus duvauceli, Cervus unicolor, Axis axis, Ovis/Capra, Panthera* sp., *Rhinoceros* sp.
Secondary consumers	*Crocodylus palaeindicus*	*Canis* sp.
Omnivores		*Sus* sp.

Crocodiles, turtles, frogs, fish and hippopotamus are a few of the representatives of Stream and Streambank community. Out of these the hippopotamus, turtles, crocodiles and frogs do spend some time on land to fulfil their physiological and other requirements. As is well known, the hippopotamus is a primary consumer while the crocodile, a secondary consumer, is an excellent scavenger.

Megaterrestrial community is the most predominant community in the Narmadā Valley. The members of this community are mostly herbivores (*Stegodon, Elephas, Bubalus*, antelopes, deer, etc.). Carnivores are few in number (*Panthera, Canis*). *Sus* is an Omnivore. Judging by this faunal compostion a diverse floral heterogenity in the valley is suggested for which climate and soil must have been very favourable. The presence of rich fossil fauna from the base of the exposed alluvium (sandy pebbly gravel) to the yellow brown silts and clayey units towards the top (below the Black Cotton Soil) suggests moist climatic conditions (with minor fluctuations) during the Middle/Late Pleistocene period. This is confimed by the examination of soil profiles in various sections and also by chemical analysis of soil sediments, which suggest moist conditions of pool environment followed by turbulent conditions of low/high intensity.

Bibos gaurus, Antilope cervicapra, and *Bubalus* sp. make best use of perennially available grasslands (which control their movements) and water resources.

Rhinoceros unicornis, though repesented by very few fossils in earlier and present collections, survives in an environment which is widely interspersed with grassland and swampy areas. The movements of rhinoceros are generally confined to the fringes of bushy, moist grassland bordering forests.

The ecological setting of cervids is represented by swampy grassland interspersed with water bodies. So far as *Cervus duvauceli* is concerned, semi-evergreen and wet evergreen forests are the habitational zones of this animal (Groves, 1982). The animal is practically dependent upon a grass cover throughout the year. According to Martin (1977) the drier condition brings *bārāsiṅghā* close to aquatic-swampy-grassy sheets. Fringes of the forests attract the animal in monsoons.

The *sāmbar* and *chītal* (*Cervus unicolor* and *Axis axis*) on the other hand, frequently traverse woods and are less frequently found in swamps. *Antilope cervicapra* and *Axis axis* can often be spotted grazing in association with cervids. Fossils of pigs are very few in the Narmadā Valley. However, one of the species, *Sus namadicus,* has been found to be a useful marker for Middle Pleistocene time bracket. Dense forests with abundant tubers, shrubs and aquatic resources were frequented by the pigs.

Elephant derives maximum energy from plant resources and is the largest consumer of vegetation. Next to fossils of hippopotamus, elephant fossils of various genera and species are found in abundance in the Narmadā Valley. The habitat of elephant comprises grassland and forest cover in equal proportions. The optimal diet of elephant appears to be a reasonable mix of browse and grass, depending upon the seasons and the types of vegetation available to it (Sukumar, 1989). Suitable ecological set-up in the Narmadā Valley from Mid to Late Pleistocene times indicates a considerable floral diversity.

Detailed habitats and ecological inferences of the genera of animals are given in the section on palaeoecological model suggested for Bhīmbeṭkā (Badam and Sathe, 1992). Ecological inferences as based on microvertebrates as also on the molluscan shells are given in appendices attached with this book.

It may be emphasized here that at Gurāriāghāṭ, where microvertebrates (*Bandicota, Gerbillus, Millardia, Tatera*) have been discovered for the first time in the Narmadā Valley, the sandy pebbly conglomeratic layer contains haphazardly oriented molluscan shells (gastropods and lamellibranchs), indicative of a channel area of a stream. The overlying and underlying pedogenically modified clays containing concretions suggest their development in flood-plain area away from streams, ponds and swamps. The palaeoecological reconstructions based on micromammals are generally more precise as they are found confined to beds of small lateral extent, deposited in a short period of time. They are sensitive to climatic changes and usually have small home ranges which in fact make them useful climatic indicators.

The total complex of the faunal remains in the Narmadā Valley, thus, indicates that the valley was a vast savannah land punctuated by flood plain lakes and swamps in which flourished a rich assemblage of hoofed mammals and reptiles. However, it may be emphasized here that there were no significant departures from the existing tropical climate in the area and therefore the valley povided a favourable ecosystem where some of the animals, which migrated from the NW because of climatic rigour in the glacial periods of the Lower Pleistocene, could survive almost up to 15,000 years ago.

The discovery of a large number of vertebrate fossils from the exposed sections on the Sher, the Umer and the Varurewa in the Devakachār area, especially near the confluence and the meanders, prompted an inquiry into the taphonomy and palaeoecology of the fossils and also about the causes for their concentration in the area. Lithological characters of various facies exposed in the trenches at Gurāriāghāṭ show that the sediments must have been laid down in a channel (gravel), inter-channel pool (gravel silt) and near-channel (silty) overbank environment. Presence of weakly to moderately developed palaeosols (?) within pool clays indicates that the beds of the Narmadā were completely dry for some time either due to lowering of water table or due to considerable lateral shifting of the river.

Distribution pattern of the fossils collected from Devakachār and the neighbouring areas during three field seasons (1992-94) under the IGRMS project suggests that thirty per cent of these, out of the total collection of fossils numbering nearly two hundred, are with fresh, unabraided surfaces. Complete fossils and intact teeth account for 20 per cent of the collection and about 50 per cent of the fossils are incomplete or fragmentary. The fossils at various localities especially in the Devakachār area, therefore, represent the death assemblage of large vertebrates which were disarticulated, fragmented and transported both before and after the burial. Since the majority of fossils are fragmentary and seem to have been transported from long distances, it is suggested that the animals must have died far away from their areas of activity. Fossils with fresh unabraided surfaces are derived generally from the near-channel overbank silty deposits and these have undoubtedly undergone least rolling. Probably animals represented by these fossils did not die far away from their areas of activity.

Taphonomically, the collection of fossils represents different types of peservation and weathering stages and shows interesting breakage patterns, various types of cracks, flakings and evidences of bacterial action. It is suggested that different taphonomic processes may have been operating differently producing different effects on death assemblages.

For example, the fossils collected from explorations generally lie scattered along the river banks of the Sher; they have been disturbed from their lithounits and have suffered heavily from fluvial activity. Most of this assemblage is supposed to be allochthonous in nature. However, the state of preservation and the distribution pattern of some fossils at Devakachār (especially those coming from the upper clayey, silty and sandy horizons) suggest that at least 30 per cent of the collection could be considered autochthonous with respect to local channels. In our excavations, we have discovered almost complete skeletons without the evidence of abrasion or transportation, of a fossil buffalo (at Bāmanghāṭ), cattle (at Gurāriāghāṭ) and buffalo (at Luknu around Devakachār) and other medium-sized mammals indicative of their complete burial/fossilization in an autochthonous situation.

Regarding the concentration of fossils in the Devakachār area, a few explanations can be put forth. It seems logical to suggest that in the Narmadā Valley, summer storms were followed by flash floods and these seasonally occurring floods may have caused panic and mass deaths of animals and their subsequent fossilization. It is well known that the Central Narmadā Valley lies in the path of cyclonic monsoon depressions which originate in the Bay of Bengal and traverse towards western and north-western India. Because of this climatic factor the Central Narmadā Valley receives frequent storm floods during the monsoons. Such floods might have been of higher frequency between 40 and 80 thousand years ago, during the Mid to Late Pleistocene times, than at present as the Narmadā at that time was in a state of aggradation compared to the present state of rejuvenation. Our taphonomic studies support this hypothesis.

Drowning, disease, starvation and fright can lead to mass deaths. The concentration of fossils at a site can also be attributed to the drought conditions when animals tend to assemble around a waterhole. As per the social behaviour of herd-oriented animals the fright of even a single animal is sufficient to cause mass panic and subsequent deaths and fossilization at a site (Behrensmeyer, 1975). However, this suggestion cannot be generalized for the entire valley and may be true only for the sites away from the main channels of the Narmadā.

The nature of the sediments and its fauna indicate that the Narmadā and its tributaries were frequently changing their courses due to overall aggradational tendency of the drainage system during the Late Pleistocene times (Badam and Sathe, 1992). The faunal evidence suggests that the vegetation during this period was of savannah type with several backswamps and water pools suitable for the animals (whose fossils are found in the valley) to thrive in the valley during the Mid-Late Pleistocene times. This and the flash flood hypothesis, may explain, in part, the causes for the concentation of fossils in certain areas of the valley. Further work on aspects like atmospheric conditions, fluvial processes, transportation and provenance of fossils needs to be taken up in detail before definite conclusions can be drawn on these aspects. Only then will we know for sure whether at certain sites, e.g. in the Devakachār area (where fossils of different animals are found together), the animals were buried together, whether they were transported together and whether they lived and died in the same habitat. The present study, however, tentatively suggests that the death, burial and transportation of animals occurred at different times and in different places in the same general area and throws better light on the taphonomy and palaeoecology of the Narmadā Valley sediments during the Mid-Late Pleistocene period than was known before.

The author has also used specimens shown in *Pl.* 5.52 to *Pl.* 5.55 for the taphonomic study of the faunal material in addition to several other fossils studied for the purpose.

8

Age of the Narmadā Alluvium

PRECISE dating of the Narmadā Alluvium has been rather problematic. It is in part due to the fact that even though a large number of Stone Age tools and fossils have been discovered from various localities in the valley from time to time, the autochthonous nature of most of the findings, especially the fossils, has been in doubt. In addition, the paucity of radiometric dates and the lack of index fossil assemblages (especially for the Middle Pleistocene) have contributed to the uncertainty of chronological framework. Furthermore, the time transgressive nature of the fossils does not allow the Lower and Middle Palaeolithic cultures, which themselves are found in a mixed context, to be placed in a proper chronological sequence.

On the basis of fossils and cultural material found in the Narmadā deposits and of attempted correlations with other Pleistocene deposits in India (Manjra, Godāvarī, Ghoḍ, Bhīmā, etc.) and elsewhere, various workers have assigned ages to the alluvium varying form Pliocene (Falconer, 1859; Theobald, 1860) to Middle Pleistocene (Pilgrim, 1905), to late Lower or early Middle Pleistocene (Zeuner, 1950), to Middle and Upper Pleistocene (de Terra and Paterson, 1939; Badam, 1979 a, b) and to Late Pleistocene (Medlicott, 1873). The underlying sandy pebbly gravel unit of the alluvium which has yielded a large number of fossils, is assigned an Upper Pleistocene age because it forms the upper part of the Older Alluvium which is considered to be Upper Pleistocene. The overlying yellow brown concretionary silt is the main fossiliferous horizon at most of the sections in the valley. Many of the fossil taxa that the Older Alluvium contains (*Cervus duvauceli, Antilope cervicapra, Bubalus bubalis, Bibos gaurus, Equus namadicus* and *Bos namadicus*) are closely related to the modern forms and hence they may not be considered to be earlier than the Upper Pleistocene. However, *Equus namadicus* and *Bos namadicus* (ancestors of *Equus hemionus* and *Bos indicus* respectively) range from Middle to Late Pleistocene and may thus no longer be considered as the index for the Middle Pleistocene (Badam, 1984). A radiocarbon date of 31750 (1770, ±

1580) BP -TF 867 (Agrawal and Kusumgar, 1975) has been obtained on fresh water molluscan shells (with half life periods of 5730 yr.) from the sandy pebbly gravel horizon (upper group of the Older Alluvium) at Ratikarar, a few km downstream of Gurāriāghāṭ in Devakachār area (Narsinghpur Distt.) assigning it to Upper Pleistocene. Another C^{14} date (25160 ± 550) BP is available from Bheḍaghāṭ (Mishra and Rajaguru, 1993)

In certain circumstances chemical assays are used to ascertain the relative dates of fossil bones within the same deposits.The values of 5.322 and 5.18 for the fluorine/phosphate ratio (100 F/P2 05) from the putative Upper Pleistocene at Devakachār is consistent with values obtained elsewhere in India and is significantly lower than ratios from fossils lower down in the same section (Kshirsagar and Badam, 1990). Furthermore, chemical studies carried out by Dr K.P. Oakley on the fossil bones from Saguṇaghāṭ and Devakachār yield 1.20 per cent and 0.97 per cent N on nitrogen tests respectively. These values, according to Oakley, are equivalent to nitrogen percentages for the Upper Pleistocene in Africa and Europe. However, there is a difference of opinion whether these values can be considered as reliable dating evidence (Badam and Grigson, 1990).

Based on palaeomagnetic dating the total Narmadā Quaternary formations are stated to be younger than 0.7 m (Agrawal et al., 1988). The hominid-bearing horizon at Hathnorā in Hoshangabad district (Sonakia, 1984) is estimated to be between 150,000 and 40,000 BP (Late Middle Pleistocene to early Upper Pleistocene) on the basis of archaeological associations (Badam, 1989). However, this date is tentative; some scholars consider the cranium to fall in the time bracket of 1,00,000 to 1,50,000 years BP (Delson, 1985) and 2,50,000 to 1,50,000 years BP (Kennedy and Chiment, 1991).

On the whole the palaeontological studies supported by the evidences enumerated above indicate that for the greater part the exposed Narmadā Alluvium belongs to the Upper Pleistocene. Middle Pleistocene exposures are found at various sections especially at Hathnorā, Barmanghāṭ, Devakachār, etc. along with the exposures of younger ages. But we need to have more radiometric dates for the alluvium in order to be able to further refine our chronological framework of the Narmadā sequence.

In this connection results of dates on volcanic ash found in several exposures of the Narmadā Alluvium are expected to provide a better understanding of the chronological framework of the Narmadā Alluvium. It may be mentioned here that an ash bed has also been traced for about 20 km by the officers of the Geological Survey of India. At Pawlāghāṭ, it occurs within the basal channel sand facies as a pinching and splitting horizon 40 to 24 cm thick. In Guruwara section, it occurs within the overbank silt as a uniformly thick (80 cm) persistent bed for about 500 m. The tentative age given for this marker is Late Pleistocene (40,000-100,000 yr. BP) and the cultural association with this formation is Middle to Upper Palaeolithic.

Though, what is written above may stand valid today till further data on chronology is available, it may be helpful to consider that as early as 1979 the author (Badam, 1979a) tried to build a relative stratigraphical time scale for the Early and Middle Palaeolithic cultures of the Central Narmadā Valley on the basis of fauna, and suggested that the Red Clay/ Boulder Conglomerate horizon which has yielded Early and Middle Palaeolithic tools along with a (?) Middle Pleistocene fauna, may be dated between 40,000 and 150,000 BP (late Middle Pleistocene to early Upper Pleistocene). The overlying cemented sandy gravel and the yellow or pink cross-bedded sand and silt (yielding Middle Palaeolithic tools along with Upper Pleistocene fauna) date from 40,000 to 15,000 BP. The topmost sub-recent black alluvium and gravel containing semi-mineralized bones is Early to Mid. Holocene. Considering the passage of time involved in the degree of post-depositional diagenetic changes and the erosional processes following them, the dating of the Lower Group of the Narmadā to the time limit of 40,000-150,000 BP appears to be within the expected range.

The upper part of the Upper Pleistocene of Narmadā (post 32,000 BP) can be correlated with the deposits of the Godāvarī, Ghoḍ and Pravara valleys from where we have numerous C^{14} dates of nearly the same range, along with identical fauna. Morphologically some of the fauna like *Hexaprotodon namadicus* from the Lower Narmadā Group (pre 32,000 BP and associated with Boulder Conglomerate) appear to be older than that recovered from the Ghoḍ, Pravara and Manjra valleys in Maharashtra.

It has also been observed that there is no micro-morphological change in the mammalian species which occur either in the Boulder Conglomerates or in the sandy pebbly gravel or at the contact between the two. It has, however, been the practice so far to date the Boulder Conglomerate horizon to the Middle Pleistocene. The present palaeontological evidence, however, considers *Sus namadicus*, *Elephas namadicus* and *Hexaprotodon namadicus* to be typical of Middle Pleistocene. Middle Pleistocene may range from 780 kyr. — 127 kyr. as per the latest estimates. Antelopes, cervids, some bovids and turtles range from late Pleistocene to Holocene and the long ranging forms (*Bos namadicus*, *Equus namadicus Elephas* spp., *Stegodon* spp., etc.) generally range from Middle to Late Pleistocene. Chronological sequence of fossils in the Narmadā Valley, as a result of the present researches, is given in Table 5.

Table 5: The Chronological sequence of fossils in the Narmadā Valley

Lower to Upper Pleistocene	*Elephas hysudricus, Stegodon insignis ganesa*
Middle to Upper Pleistocene	*Elephas namadicus, Sus namadicus, Equus namadicus, Bos namadicus*
Middle Pleistocene*	*Elephas namadicus, Sus namadicus, Hexaprotodon namadicus*
Upper Pleistocene*	*Sus palaeindicus, Hexaprotodon palaeindicus*
Upper Pleistocene to Holocene	*Antilope cervicapra, Cervus unicolor, Cervus duvauceli, Axis axis, Boselephus tragocamelus Equus asinus, Elephas maximus*

* Index fossils

Notes

1. The above biochronological sequence, which revises our earlier concepts, is based on our extensive work in Son, Belan and Narmadā valleys and valleys of Western Maharashtra including Godāvarī, Manjra, Kṛṣṇā, Ghoḍ, etc.
2. It has not been possible to include here the latest data regarding chronology of the Narmadā Alluvium and the several dates that have been obtained by various methods. However these will form the subject matter of a future publication as mentioned elsewhere.

9

Concluding Remarks

Study at a Glance

DETAILED explorations and palaeontological excavations were carried out in the Central Narmadā Valley, Dist. Narsinghpur (M.P.) from 1992 to 1994 as part of the Project "Taphonomy and Palaeoecology of the Fossil Vertebrates from the Central Narmadā Valley, Dist. Narsinghpur, M.P.," sponsored by the Indira Gandhi Rashtriya Manav Sangrahalaya, Bhopal. The fieldwork was conducted under the direction of Dr G.L. Badam, palaeontologist, formerly of Deccan College, Pune. The co-investigator of the project was Shri D.K. Jain (an anthropologist from IGRMS, Bhopal). The present book is the outcome of the work carried out under the above project.

In view of the abundance of fairly preserved geological sections as well as fossils at Devakachar, Dist. Narsinghpur, excavations were mainly carried out in this locality. A pilot trench and several trial trenches were dug here to look at the sedimentological context of fossil assemblages as well as its bearing on the preservational aspect of the death assemblages during the Pleistocene period. On the whole, the pilot trench and trial trenches yielded vertebrate fossils, molluscan shells, prehistoric artefacts and also provided useful geomorphological and sedimentological data to suggest that the pattern of sedimentation was un-uniform in the shallow stagnant water basin, followed by flash floods as evidenced by the grain size of sub-recent alluvium.

In addition to the collections made previously, a large number of skeletal and cranio-dental elements of *Elephas hysudricus*, *Elephas namadicus*, *Hexaprotodon namadicus*, *H. palaeindicus*, *Bos namadicus*, *Bubalus palaeindicus*, *Bibos gaurus*, *Antilope cervicapra*, *Equus namadicus*, *Crocodylus palaeindicus* and *Trionyx* sp. were collected from these excavations and from explorations in and around Devakachar. However, one of the important discoveries made during the project work, and for the first time in the valley, is the record of a microvertebrate community found in late Pleistocene sediments at Devakachar (see Appendix I). The fossil remains were found concentrated in the top gravelly bed which is overlain by sub-recent yellow silt followed by Black Cotton Soil. The findings comprise partial skulls, isolated teeth, incisors, limb bones, vertebrae, etc. referable to the genera *Tatera, Gerbillus, Bandicota* and *Millardia.* A few fish dentaries and crocodilian teeth were also found from this gravelly horizon. The faunal collection was subjected to a detailed taphonomic study.

The discovery of micromammals helped us throw precise light on palaeoecology of the Narmadā sediments, hitherto derived mainly on the basis of megavertebrates. The fossil remains of murids (*Millardia, Bandicota*) are very rare and the present discovery has a great bearing on the taxonomic and phylogenetic reconstruction of the Pleistocene murids in general. So far, the murid remains in the subcontinent have been discovered only in the Upper Siwaliks of India and Pakistan and the Karewas of Kashmir, dated from late Pliocene to Early Pleistocene. As regards the evolutionary relationships of these genera with their predecessors in the NW, it is necessary to have numerically several complete specimens from different chronologically controlled geological strata.

Rodents are confined to specific habitats and have life span of relatively short duration with small home ranges. As a result, their occurrence, confined to one particular horizon, indicates the areas of preference with respect to the local ecology of the area. The skeletal elements are in a fairly good state of preservation and their ratio clearly suggests that the animals have died at the spot of biocoene activity. On the whole, this is the most important contribution of the fieldwork carried out under the present project in the Central Narmadā Valley.

An interesting worked bone was found in association with the fossil rodents in the top gravel bed at Devakachar. The upper part of the bone surface has been polished and two longitudinal chips have been removed, possibly in an attempt to make a chisel from the bone fragment. Texture and colour of the worked area suggest this to be a pre-fossilization activity. If the presence of the bone tools is confirmed by more data, the antiquity of bone tool making in India can be pushed back in time. So far, the earliest report of bone tool making is known from the Kurnool Caves in Andhra Pradesh, dated to 25,000 BP.

At Devakachar, a thin deposit of volcanic ash has been located within a non-*in situ* gravel bed, about 5 m in the top portion of a section at Sūkhā Nālā. Radiometrically, the exposed Quaternary sediments in the Narmadā Valley around here have been dated to 31,750 yrs. BP and at Bheḍaghāṭ to 25,160 yrs. BP. In fact, the results need to be substantiated by more dates with the help of different physical methods so that the alluvium is put in a proper chronological framework. Against this background, the absolute dating of the volcanic ash is important as it is expected to provide us with a definite geochronological record of Pleistocene sediments in the Narmadā. Previously, volcanic ash had been discovered at Bori (dated as Early Middle Pleistocene) in Kukdi river valley (Dist. Pune, Maharashtra) by the Deccan College team and in the Son Valley in Madhya Pradesh by Geological Survey of India. Recently, volcanic ash has also been discovered in Purna Valley near Akola and at Morgaon in the Bhima Valley, both in Maharashtra, which is being jointly investigated by the scientists from the Deccan College, Pune, University of Poona, and the Physical Research Laboratory, Ahmedabad. The results will help us to date Quaternary sediments of the Peninsular India comprehensively and this may turn out to be the most outstanding contribution on Quaternary geology in recent times.

The recent fieldwork has also yielded a large number of Stone Age and Mesolithic tools from stratified context and sometimes in association with fossils, from the top of gravel beds in proximity to main channels of the Narmadā and its tributaries. Typologically, the tools can be classified into different types of flakes, scrapers, cores, blades, points, etc.

Explorations were also conducted at other localities like Ratikarar, Jhānsīghāṭ, Talayaghāṭ, Guarīghāṭ, Barmanghāṭ, Bheḍaghāṭ, Sūrighāṭ, Pawla, Hoshangabad, etc. and several fossils, Stone Age tools and the relevant geological data were collected. A large number of molluscan shells of lamellibranchs and gastropods found associated with gravelly horizons indicate fresh water deposition of the sediments and may give better clues to the chronology and palaeoecology of the Narmadā sediments (see Appendix II). The pedological studies of the various soil samples have been taken up in order to understand the environmental factors operating during the building up of the alluvium which has been deposited fairly rapidly without any significant time gap. From the soil chemical analysis carried out on samples from two profiles in the Devakachar area, it can be suggested that these sediments had their source in an area where the rainfall appears to have been significantly high as indicated by pH (neutral to slightly acidic), low silica content and fairly high Fe_2O_3 content (see Appendix III).

To sum up there may be some lacunae in our understanding of the Narmadā alluvium in its totality at the present stage of our knowledge but we have tried to tackle the problems of chronolgy, taphonomy and palaeoecology from a multidisciplinary point of view with some success. Future work in the valley using more scientific techniques may unfold more mysteries that lie buried in the alluvium.

In the end, we may like to mention that a lot of new work on various aspects of Central Narmadā Valley, e.g. new chronological data, flood history and river behaviour has been done during the last five years. However, none of it is mentioned here in great detail, as it would disturb the scope and format of the work, which is essentially restricted to studies of Quaternary palaeontology and related aspects of the Narsinghpur District. However, a detailed study and interpretation of the new data is planned for a future publication.

Conclusions

THE following tentative conclusions can be drawn from a multidisciplinary study of the project in the Narmadā Valley:

1. The Narmadā Valley has assumed much significance with the discovery of a partial skull of a hominid, *Homo erectus narmadiensis* at Hathnorā (Hoshangabad Distt.) in 1984 by the officers of the Geological Survey of India. Though work on this valley has been carried out for more than 150 years by several foreign and Indian scholars including the present writer, the hominid discovery has infused renewed interest in the study of various aspects of the valley.
2. The recent and the first discovery of microvertebrates in the valley is expected to interest several microfaunal experts because of their importance as chronological and palaeoecological markers. The assemblage reported from the Narmadā indicates an early emergence of these modern rodents in the Indian subcontinent.
3. The mega fossil collection from the Central Narmadā Valley is mostly dominated by large animals like *Elephas hysudricus, Elephas namadicus, Segodon insignis ganesa, Hexaprotodon namadicus, Hexprotodon palaeindicus, Rhinoceros unicornis, Bibos gaurus, Bos namadicus, Bubalus palaeindicus, Equus namadicus,* etc. The occurrence of medium-sized herbivores like *Axis axis, Antelope cervicapra, Cervus unicolor, Cervus duvauceli, Boselaphus tragocamelus,* etc. is also common. The reptilian species are represented by crocodiles and turtles and the birds by ostrich egg shells at several sites in the Narmadā Valley. Fish vertebrae and frog dentaries have also been collected.

4. The Pleistocene environment in the Central Narmadā Valley supported the above diverse fauna, suggesting a diversity of ecological niches, which was dominated by dense forests, open grasslands and a healthy fluvial system. The fluvial environment was of high energy type, capable of transporting big and heavy bones and boulders. Paucity of small vertebrates, concentration of large vertebrates at meanders and gulleys away from the main channels of the Narmadā, are indirect effects of huge flash floods caused as a result of channelling of monsoon winds from the Bay of Bengal to the central part of India.
5. The taphonomic signatures on the fossils also hint at the dominance of the fluvial system which might have occurred spasmodically resulting in large-scale deaths and subsequent burial of animals. The fossils recovered from the lag deposits are invariably well preserved depicting their immediate burial. However, the fossils are rarely found in the form of entire skeleton, suggesting that some dispersal has taken place even with the well preserved ones. Though most of the fossils are not autochthonous to the localities of their find, yet some can be identified as being *in situ* with respect to local channels. These were either not transported at all or transported to a small distance before and even after their burial.
6. The density of Hippopotamus suggests the prevalence of swampy pool environment and also supports its wide distribution in the Indian subcontinent during the Pleistocene period before becoming extinct.
7. *Sus namadicus, Hexaprotodon namadicus* and *Elephas namadicus* can be considered as the most reliable index fossils for the Middle Pleistocene time bracket.
8. A geological and palaeontological study of the Central Narmadā Valley over the last few decades indicates that the Narmadā and its tributaries frequently changed course due to an overall aggradational tendency of the drainage system during Mid to Late Pleistocene times. The Central Narmadā, because of its geographical location, did receive comparatively large amounts of rainfall. This climatic condition led to frequent floods in the aggrading Narmadā and its tributaries. In fact the processes of the Narmadā are controlled by high magnitude floods, the seasonal pattern of the Indian monsoon and the tectonics of the basin.

9. The Narmadā was rejuvenated in the early Holocene. The groundwater was reduced to modern levels which resulted in a number of swamps and backswamps in this environment. At the same time, the early Holocene was characterized by increased rainfall and vegetation density, resulting in the receding of open spaces. This provided an ideal habitat for herbivorous ungulates, proboscideans and reptiles to thrive in the area. The carnivores and other animals were probably concentrated in the forests near the banks of the rivers. Therefore, the overall picture appears as though the habitat of the varied Pleistocene animals encompassed a large geographical region in and around the Central Narmadā Valley.
10. A paper "Animal Depictions of Rock Art and Palaeoecology — A case study at Bhīmbeṭkā, M.P., India" published in 1992 has been reproduced here with minor revision (see Appendix IV) because of its detailed ecological/palaeoecological treatment of the animals found in rock paintings of Bhīmbeṭkā and their extensions into past geological forrmations as in the Narmadā Pleistocene deposits.
11. On faunal and chronological grounds, most of the ancient alluvium of the Narmadā Valley can be placed in Middle–Late Pleistocene time bracket. However, some evidences of lower Pleistocene data of some sections of the Narmadā have also come to light recently.
12. Distribution of fauna in the Pleistocene deposits of India and their probable ages is given in Table 7.
13. Quaternary Biological and chrono-stratigraphy in India is compiled in Table 8 to show the status of the Narmadā Valley in relation to the Indian stratigraphy.

Table-7

Distribution of Fauna in the Pleistocene Deposits of India and their Probable Ages[1].

Locality	Important Fauna	Stone Age Tools	Associated Cultures	Probable Age	Environment
Kurnool, Ghod, Manjra, Pravara, Godavari	*Canis* sp., *Bubalus* sp., *Cervus* sp., *Bos namadicus,* *Elephas hysudricus, Elephas maximus, Rhinoceros unicomis, Bos indicus, Hexa-protodon palaeindicus.*	Bone tools, Burins, Blades, Points. Scrapers, Flakes, Blades, Points, Borers.	Upper Palaeolithic Late Middle Palaeolithic	Late Upper Pleisto-cene (also dated by C-14; 40,000 to 15,000 B.P.)	Savannah type with pockets of forests and swamps (humid in Kurnool).
Central Narmada (Upper Group); Paimar	*Equus namadicus, Bos namadicus, Hexa-protodon palaeindicus, Elephas hysudricus, Stegodon insignis-ganesa, Cervus* sp.	Scrapers, Flakes, Flake-blades, Points, Borers, Handaxes, Clea-vers, Polyhedrals, Discoids, Choppers.	Middle Palaeolithic (Late Acheulian)	Early Upper Pleistocene.	Savanna grassland interspersed with swamps
Central Narmada (Lower group)	*Equus namadicus, Bos namadicus, Hexa-protodon namadicus, Sus namadicus, Elephas hysudricus, Stegodon insignis ganesa.*	Choppers. Handaxes, Cleavers Flakes	Lower Palaeolithic (Acheulian)	Middle Pleistocene	Savanna grassland interspersed with swamps
Lower Karewa, Pinjor of Upper Siwaliks	*Equus sivalensis, Rhinoceros sivalensis, Rhinoceros palaeindicus, Bos acutifrons, Archidiskodon planifrons, Sivatherium giganteum.*	No authentic or *in situ* Stone Age Tools[2] *Elephas hysudricus*		Lower Pleistocene	Valley and open savanna grassland with lakes and swamps.

Note:

1. This table is updated till 2000.
2. Although there have been some reports of cultural findings recently, these must be confirmed before their presence is accepted.

Table-8

Quaternary Bio-Cultural and Chrono-Stratigraphy in India

Geo Time Scale	Archeo Time Scale	Approx Age (BP)	Significant Cultural Traits	Fauna	Dominant Climate
Late Holocene	Historical (Iron Age)	3000 B.P. onwards	Writing, Iron technology and Advanced agriculture known	Modern animals	Good monsoons prior to CE 1000, Bad monsoons, between 14th and 17th C.A.D.
Mid Holocene	Chalcolithic/ Neolithic	4500-3000 B.P.	Use of copper and stone, Beginning of agriculture, Advanced town planning, Knowledge of writing, Polished tool technology	Domesticated and wild animals	Weak monsoons around 3000 B.P.
Early Holocene	Mesolithic	10-12,000 to 4,000 B.P.	Composite tool-technology, Pastoralism and fishing dominant	Beginning of domestication of animals	Very good monsoons (8000-4000 B.P.), Winter rains strong.
Terminal Pleistocene	Upper Palaeolithic	40,000 to 10-12000 B.P.	Palaeolithic art and blade tool technology, Hunting and food gathering	*Equus hemionus*, *Bos namadicus*, *Rhinoceros* sp., *Bubalus* sp., *Elephas maximus*, *Cervus* sp., etc.	Poor monsoons.
Late Pleistocene	Middle Palaeolithic	130,000 to 40,000 B.P.	Hunting and food gathering, use of fine grained chalcedony, chert etc. for making scrapers and flake tools, use of quartzite and basalt continues.	Same as above + *Equus namadicus*, *Bos namadicus*, *Hexaprotodon* sp., *Elephas hysudricus*, *Sus* sp., etc.	Good monsoons around 120,000 B.P., Moderate to weak monsoons from 75,000 Yrs. onwards.
Middle Pleistocene	Lower Palaeolithic (Early to Late Acheulian)	700,000-130,000 B.P.	Hunting and food gathering, use of quartzite, basalt for making handaxes, scrapers, cleavers, flakes, choppers etc.	Same as above + *Hexaprotodon namadicus*, *Sus namadicus* etc.	Fluctuating monsoons, Strong to weak.

Lower Pleistocene	Suspected Lower Palaeolithic in Pakistan (Peshawar) and India (Maharashtra)	2.0 My-0.7 My (Peshawar : 2 My, Bori : 670,000 Yrs.)	Hunting and food gathering, use of basalt for making flakes and Choppers	*Equus sivalensis*, *Bos acutifrons*, *Rhinoceros sivalensis*, *Hexaprotodon sivalensis*, *Stegodon insignis ganesa*, *Archidiskodon planifrons*, etc.	Fluctuating monsoons, Strong to weak.
Plio-Pleistocene		2.47 My-1.87 My		First Appearance of *Equus* and Extinction of *Hipparion*	Transition from tropical humid climate to seasonal monsoonic climate.
Pliocene		5 My-2.5 My		*Hipparion* predominant	Tropical humid climate
Miocene		20 My-5 My			

Table compiled in 2002 by G.L. Badam and S.N. Rajaguru, Deccan College Post-Graduate and Research Institute, (Deemed University), Pune.

Notes:

C 14 Dates upto 40,000 BP
K/AR Dates for period 1 My and beyond

Appendix I
Microvertebrates

Rajeev Patnaik

Introduction

OVER the last hundred years, focused research on fossilized remains of large vertebrates of the Narmadā valley deposits of Central India has yielded very interesting results. One of the most important finds which has brought the Narmadā Valley into limelight is the discovery of a hominid skull (Sonakia, 1984; Lumeley and Sonakia, 1985). Besides the skull, a great wealth of large vertebrates and Palaeolithic tools have been unearthed from Narmadā Valley deposits in the last two centuries (Badam, 1984). In recent years, data regarding microvertebrates such as rodents, insectivores, fish and frogs (Patnaik et al., 1995; Patnaik, 1995) and a bird (Patnaik and Sahni, 1995) have also been obtained from these sediments. New results have emerged regarding the age (Joshi and Kshirsagar, 1986; Agarwal et al., 1988; Sonakia, 1998) and sedimentology (Khan, 1984; Khan and Sonakia, 1992) of the Narmadā Quaternary deposits.

In this context, a preliminary report on the microfossils (discovered for the first time) which include ostracodes, *Candona fabaeformis, Darwinula stevensi, Ilyocypris bradyi* and *Cypridopsis* sp. and molluscs belonging to *Hydrobiidae* indet., cf. *Gyraulus, Pisididae* indet. is made here (*fig.* 3). The present assemblage has common elements in the Karewas of Kashmir and Upper Śiwāliks (Bhatia et al., 1985; Bhatia, 1969; Bhatia and Mathur 1973). The interpretation based on microvertebrates recovered (Table 1) earlier (Patnaik et al., 1995; Patnaik 1995) are improved herein. The incisors of *bandicota* and *Tatera* have been subjected to scanning electron microscopic studies in order to gain knowledge about the microstructural reinforcement of enamel. The taphonomic and palaeoecologic interpretations based on micromammals have been reviewed. The distribution and phylogenetic relationships of small mammals of the Narmada Valley in space and time has been discussed.

Tooth enamel is made up of hydroxyapatite crystallites. These thread like crystallites are arranged parallel to each other to form bundles called prisms. The inter-prismatic space is also occupied by these crystallites. The prisms start from the enamel dentine junction and end up on the outer surface of the enamel. The incisor enamel is divided into outer radial and inner Hunter-Schreger Bands (*fig.* 7 A-B). There are basically three types of HSBs; the one with 1-2 band thickness is called uniserial, that with 2-4 band thick is called pauciserial and those with more than 4 band are termed as multiserial. The Narmadā rodents show typical unserial pattern.

Material and Methods

The microfaunal assemblages were recovered by excavating the mollusc-bearing conglomeratic layer at Devakachār and mudstone layer at Hathnorā (*figs.* 1 and 2). A small portion of the layer was subjected to dry and wet sieving in the field. In this paper, only teeth are described and compared.

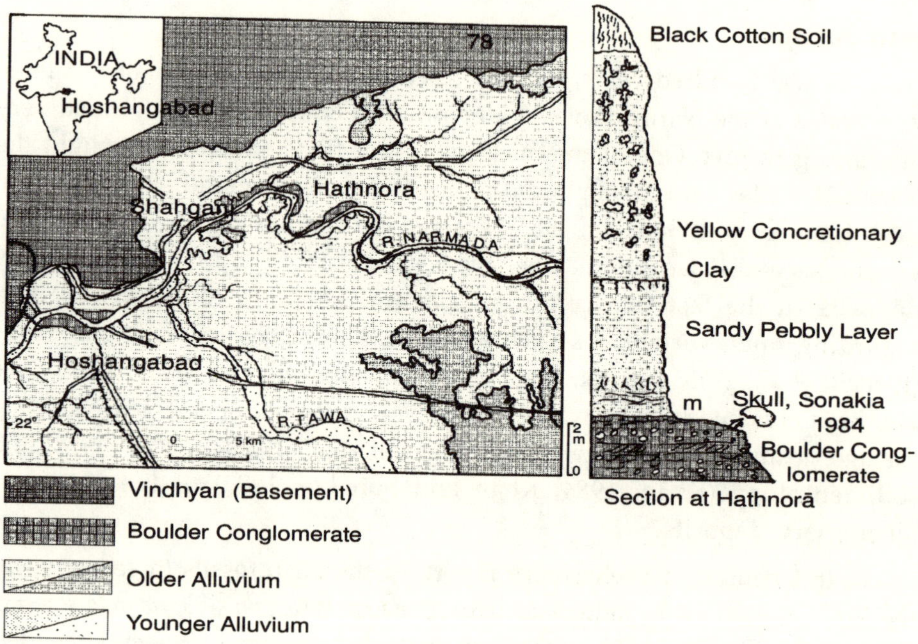

fig. 1. Geological map of the area and lithocolumn at Hathnorā (Modified after, Khan and Sonakia, 1992), m, microfossil yielding layer.

Dental terminology for murids and cricetids proposed by Jacobs (1978) and Tong (1986) respectively are followed (*figs.* 4 and 9). For comparison, recent material housed at the Zoological State Collection, Munich, the Koenigs Museum, Bonn (Germany), the Zoological Survey of India collection, Calcutta and the Bombay Museum of Natural History, Mumbai have been used. Measurements were made using a microscope fitted with a micrometre and sketches were made using a Camera Lucida fitted on a Wild M3 microscope.

For enamel ultrastructure studies, the incisors were embedded in polyster resin and cut longitudinally transversely. The sections were polished using 40 and 80 mesh carborundum powder. The polished surface was etched for

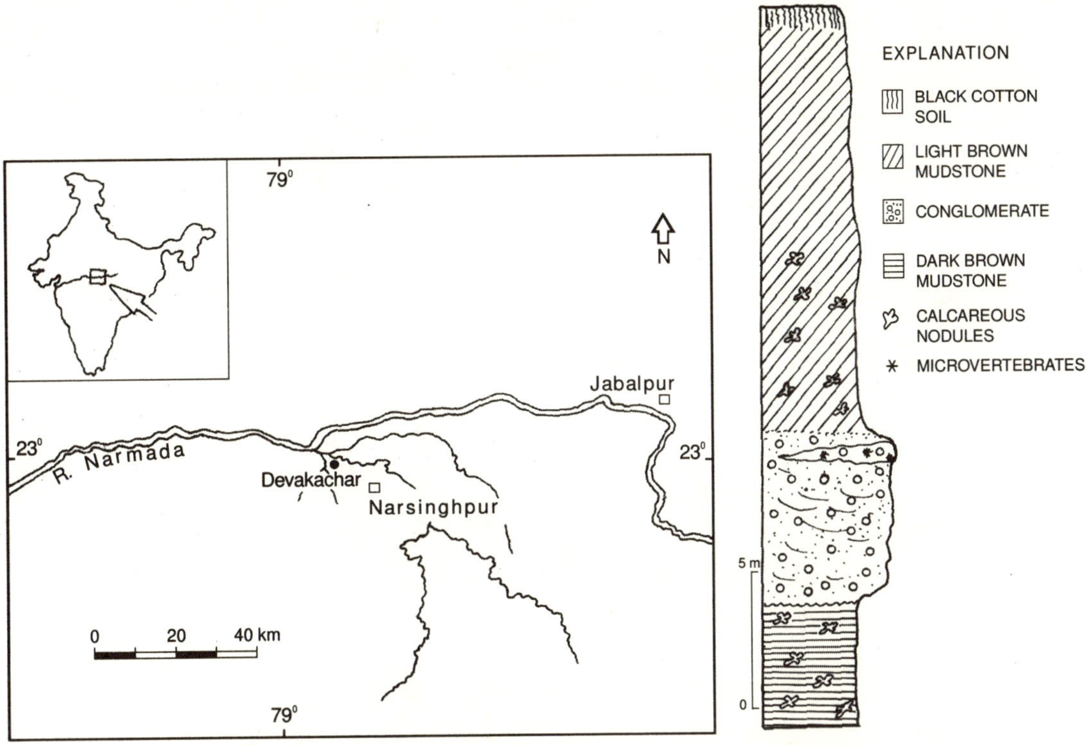

fig. 2. Locality map (A) and lithocolumn (B) at Devakachār (Patnaik, 1995).

20 seconds using 5 per cent HCL and was cleaned using an ultrasonic cleaner. The surface was mounted on a stub and coated with gold sputter. Then the sample was viewed and photographed using a scanning electron microscope.

Class	:	Mammalia
Order	:	Rodentia
Family	:	Muridae

Murids (rats and mice) are considered to be among the most successful groups of living mammals. They have adapted themselves to varied ecological conditions and show marked species diversity. Murids are believed to have originated in the Indian subcontinent around 15 myr ago in the form of *Antemus* (Jacobs, 1978). To date, a number of fossil Murids are known from Africa, China, Afghanistan, Pakistan and India. They have also made their way to Europe. Neogene Śiwālik sediments have yielded numerous Murid specimens (Jacobs, 1978; Patnaik, in press). Apart from the Śiwāliks, murids are also known from the Karewas of Kashmir (Kotlia, 1992). Śiwālik muridae include *Antemus, Progonomys, Karnimata, Parapelomys, Golunda, Mus, Hadromys, Cremnomys Bandicota, Nesokia,* cf. *Millardia* and cf. *Lemniscomys. Mus* and *Apodemus* are known from the Karewas of Kashmir. Murid molars show great variation in size, shape and the position of cusps, and can be identified up to the specific level based on molar morphology (Misonne, 1969; Jacobs, 1978).

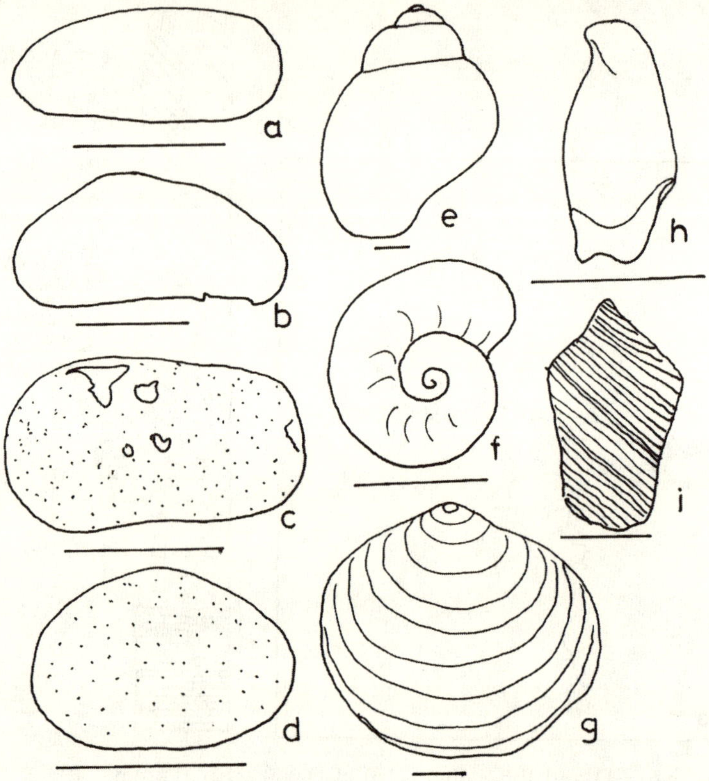

fig. 3 **a.** *Darwinula stevensi* (VPL/RP-M1); **b.** *Candona fabaeformis* (VPL/RP-M2); **c.** *Ilyocypris bradyi* (VPL/RP-M3); **d.** *Cypridopsis* sp. (VPL/RP-M4); **e.** *Hydrobiidae* indet. (VPL/RP-M5); **f.** Cf. *Gyraulus* sp (VPL/RP-M6); **g.** *Pisididae* indet. (VPL/RP-M7); **h & i.** isolated tooth and scale of *Cyprinidae* indet (VPL/RP-M8&9). Bars represent 5mm.

Genus — *Millardia* Thomas, 1911

Millardia cf. *M. meltada*

Referred material: PLDC-M numbers, maxillae (1, 2, 3, 4),

Mandibles (5, 6, 7, 8, 9)

Locality : Devakachār

Stratigraphic range : Devakachār Formation

Age : Upper Pleistocene

DESCRIPTION

M1 comprises three cheverons with anterostyle, lingual and labial anterocones in the first cheveron, protocone and paracone in the second, and hypocone and metacone in the third. The anterostyle is large, subrounded and situated relatively posterior to the labial and lingual anterocones. The anterostyle is strongly connected to the lingual anterocone. The lingual anterocone is a large cusp, gently inclined and moderately connected to the labial anterocone,

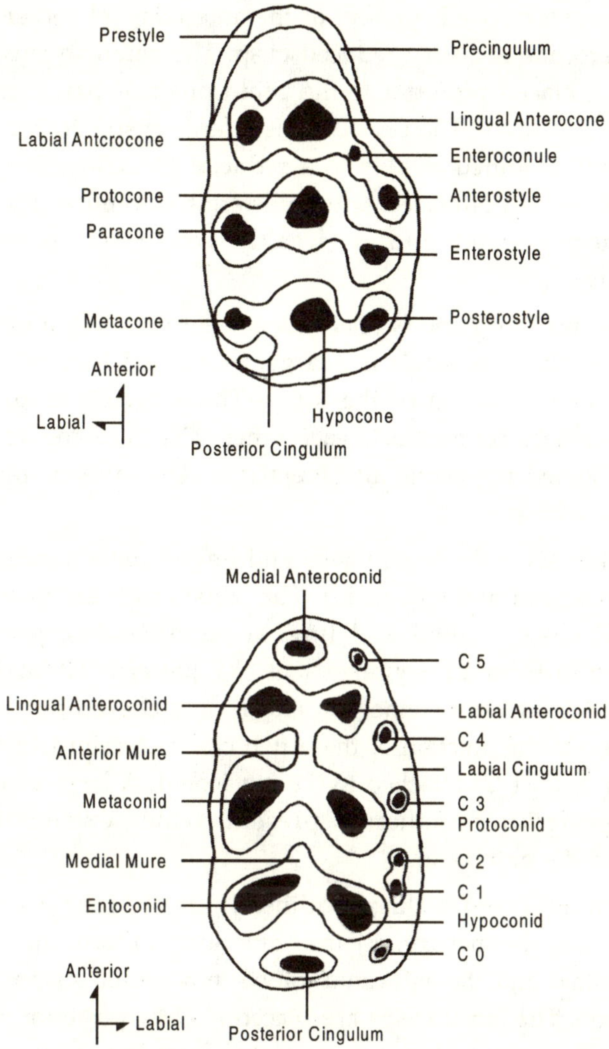

fig. 4. Dental terminology of murid first upper and lower molars (after Jacobs, 1978).

through a minor cusp. The labial anterocone is reduced, anteroposteriorly compressed and situated slightly anterior to the anterostyle and posterior to the lingual anterocone. On the second cheveron, the enterostyle is large and sub-triangular in shape and is placed relatively posterior to the protocone and paracone. The protocone is large and joined laterally to the paracone through a minor cusp. The paracone is moderately built, anteriorly directed, anterior to the enterostyle and posterior to the protocone. The hypocone is the largest cusp on the molar connected and situated posterior to the metacone. The metacone is a small cusp and is joined to the paracone by a thin connection (stephanodonty).

M2 comprises two cheverons with the enterostyle, protocone and paracone forming the first, and the hypocone and metacone forming the second. A large anterostyle is present at the anterolingual corner of the tooth. The labial anterocone is a very reduced cusp. The enterostyle is a large subtriangular cusp, placed posterior to the protocone and paracone, and joined strongly to the protocone. The protocone is a large cusp situated at the centre of the molar and is joined weakly to a reduced paracone. The hypocone is again a large cusp and is placed at the posterior extremity. The metacone is reduced, almond-shaped and is weakly connected to the paracone (stephanodonty).

M3 is wider than long and is relatively smaller than M2. It has a triangular occlusal outline. The anterostyle is a large, round and isolated cusp situated at the anterolingual margin of the tooth. The enterostyle, protocone and paracone are strongly connected to each other. The third cheveron comprises very strongly joined hypocone and metacone. The latter is placed slightly anterior to the former.

M1 comprises six main cusps (labial and lingual anteroconids, protoconid, metaconid, hypoconid and entoconid). The labial cusps are posterior relative to the lingual cusps. Labial and lingual anteroconids, protoconid and metaconid join to form an asymmetrical "X" pattern. The hypoconid and entoconid share an anterior connection and form a cheveron. A small "mure" can also be seen at the junction of these two cusps. A faint labial cingulum is present with a distinct C1 attached to the metaconid. A large anteroposteriorly compressed posterior cingulum is present, which extends between the hypoconid and the entoconid.

M2 has a trapeziform occlusal outline. There are four main cusps, with the protoconid and the metaconid joined anteriorly to form the first cheveron, and the hypoconid and the entoconid joined in a similar manner to form the second cheveron. The labial cusps are placed slightly posterior relative to the lingual cusps. The entoconid is well separated from the metaconid. The labial anteroconid is a small cusp at the anterolabial border of the tooth and is attached to the protoconid by a moderately weak connection. A very small isolated C1 is present on the faint labial cingulum. The posterior cingulum is anteroposteriorly compressed and extends along the posterior margin of the tooth between the hypoconid and the entoconid.

M3 is a relatively smaller molar with a triangular occlusal outline. The protoconid is very strongly connected to the metaconid and forms almost a straight cheveron on the anterior portion of the tooth. The hypoconid is a large arcuate and situated near the labial margin of the tooth (*fig.* 5).

COMPARISONS AND DISCUSSION

The present molars most closely resemble those of *Millardia meltada* (extant taxon). A few differences noted are the presence of stephanodonty on M1, M2 and minor cusps between the anterostyle and the lingual anterocone and the enterostyle and the protocone in the present specimens. *Millardia gleadowi* (extant taxon) differs from *Millardia* cf. *M. meltada* in having an M1 with

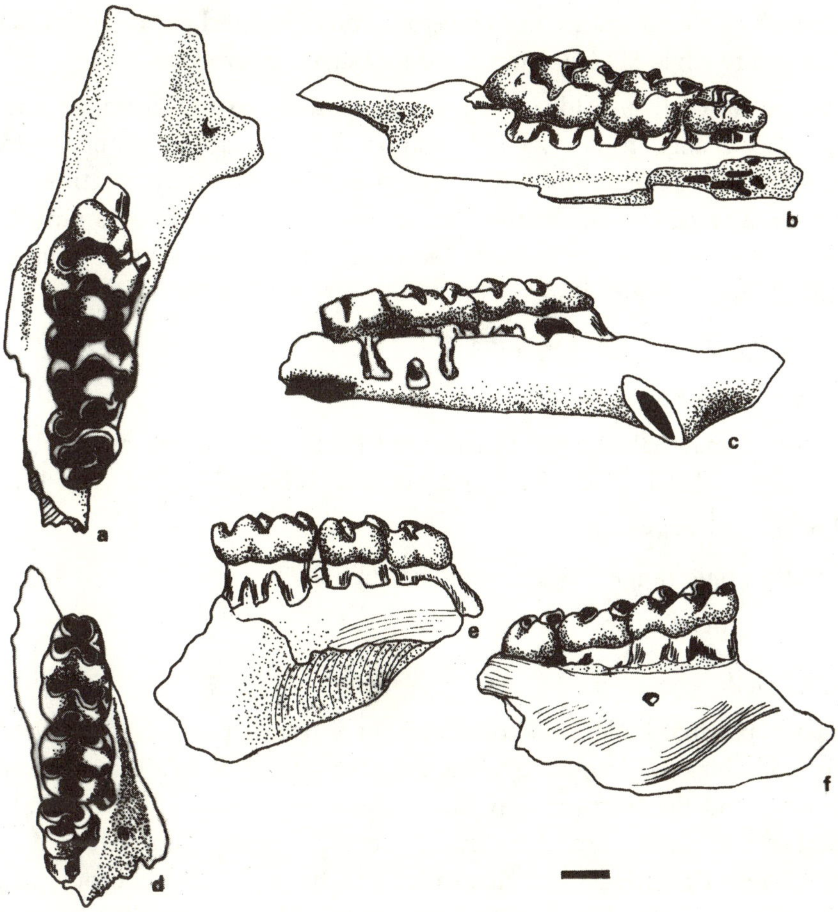

fig. 5. *Millardia* cf. *M. meltada.* **a.** occlusal; **b.** lingual and **c.** labial views of maxilla (PLDC-M-1); **d.** occlusal; **e.** lingual and **f.** labial views of mandible (PLDC-M-5). Bar represents 1 mm.

anteroposteriorly compressed and posteriorly placed anterostyle and lacking stephanodonty. The difference betwen *Millardia kathleenae* and *Millardia* cf. *M. meltada* is that in the former M1 has the lingual anterocone more anteriorly elongated, the anterostyle laterally compressed and more posteriorly placed compared to that seen on *Millardia* cf. *M. meltada.* The lower molars of *M. kathleenae* lack the labial cingulum and M1 shows the protoconid detached from the metaconid, and the lingual and labial anteroconids.

The two fossil finds, i.e. *M. coppensi* and *M. taiebi* from the Late Pliocene, Hadar Formation of Ethiopia (Sabatier, 1982) differ considerably from *Millardia* cf. *M. meltada.* M1 of *M. coppensi* has a relatively posteriorly placed and laterally compressed anterostyle and a metacone which is larger in size and anteriorly placed. On M1 of *M. taiebi,* the lingual anterocone is elongated more anteriorly and surmounted with a prestyle while the anterostyle is laterally compressed and posteriorly placed relative to those on M1 of *Millardia* cf. *M. meltada. Millardia* of the Talbot formation differs from *M. meltada* in having an M1 with an almost straight first cheveron, a much reduced paracone and lacking stephanodonty while M2 has a much reduced labial anterocone.

As we move from older *Millardia* specimens to the younger ones, a considerable change in the size and position of the metacone can be observed. With time, the metacone is reduced and moves anteriorly relative to the hypocone. The labial anterocone also shows a reduction in size with time. The lower molars of *Millardia* (from the examples studied) do not exhibit any notable change through time.

Genus - *Bandicota* GRAY, 1873

Bandicota cf. *B. bengalensis*

Referred material : one skull fragment (PLDC-M-10), one mandible with M1 (PLDC-M-11) and M2 and isolated M1 (PLDC-M-12).

Locality : Devakachār

Stratigraphic range : Devakachār Formation

Age : Upper Pleistocene

DESCRIPTION

M1 is a hypsodont molar and comprises three transverse cheverons; the first consists of labial and lingual anteroconids, the second of an protoconid and a metaconid and the third of a hypoconid and an entoconid. The labial and lingual anteroconids are strongly connected (highly laminated) and an incompletely developed anterior mure can be seen. The protoconid and metaconid, and the hypoconid and entoconid are strongly connected respectively to form well laminated cheverons. M1 lacks both posterior and labial cinguli.

M2 is almost rectangular in occlusal outline. The labial anteroconid is situated at the anterolabial corner of the tooth and is strongly connected to the protoconid which is strongly joined to the metoconid to form a well laminated first cheveron. The hypoconid and entoconid are also connected very strongly to form the second cheveron (*fig*. 6).

Incisor enamel clearly shows presence of outer radial enamel and inner Hunter-Schreger Bands (*fig*. 7 B-E). The prisms of radial enamel run parallel to each other and are pointed towards the tip of the incisor. The Hunter-Schreger Bands are typical uniserial type (1-2 band width) and are made up of alternatively decussating prisms. Crystallites of interprismatic matrix are perpendicular to prisms.

COMPARISONS AND DISCUSSION

The two extant forms of *Bandicota*, *B. indica* and *B. bengalensis* differ from each other in the degree of lamination and the position of the cusps on upper and lower molars. *B. bengalensis* has more laminated lower molars (labial cusps are attached strongly to lingual cusps) and lack a posterior cingulum.

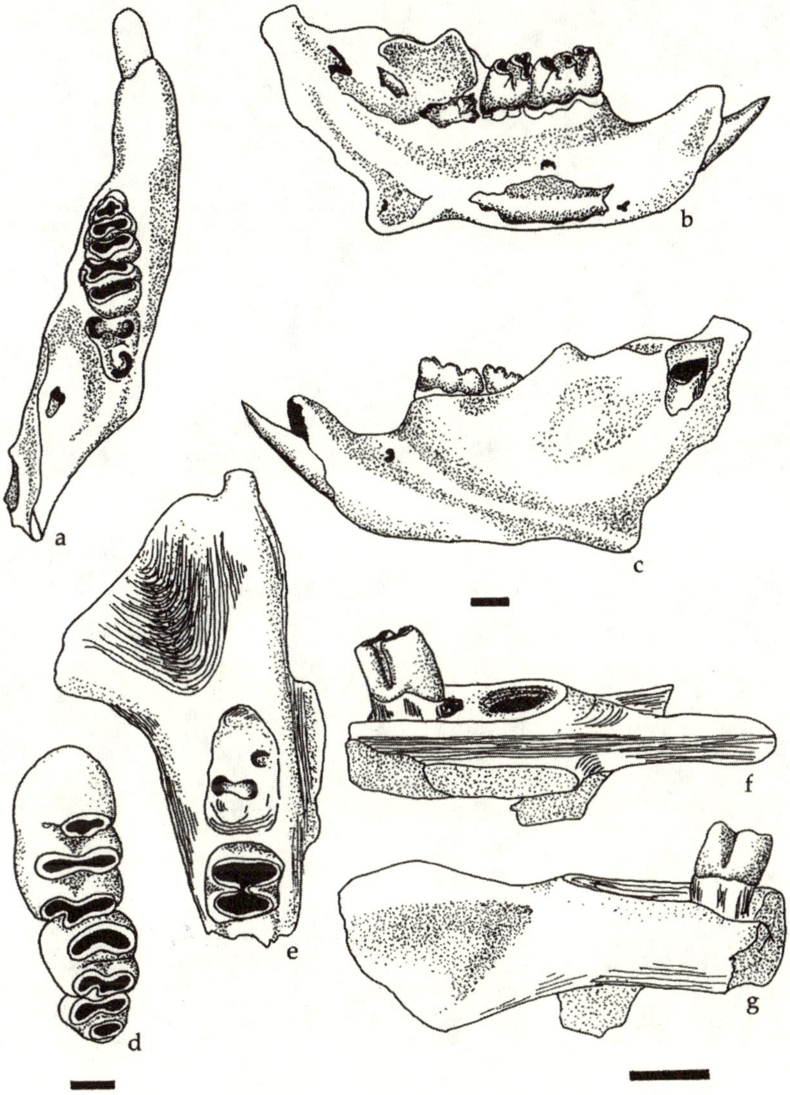

fig. 6. *Bandicota* cf. *B. bengalensis.* **a.** occlusal; **b.** lingual and **c.** labial views of mandible (PLDC-M-11), bar represents 2 mm. *Tatera* cf. *T. indica;* **d.** occlusal view of maxilla (PLDC-G-1), bar represents 1 mm. *Gerbillus* sp.; **e.** occlusal; **f.** lingual; and **g.** labial view of maxilla with M (PLDC-G-2). Bar represents 2 mm.

All the features on M1 and M2 of *B.* cf. *B. bengalensis* resemble those seen on M1, M2 of *B. bengalensis*. Besides the molars, the skull fragment of *B.* cf. *B. bengalensis* is strikingly similar to that of a *Bondicota bengalensis* in dimension and position of the various sutures (*fig.* 12 a). Another closely related taxa is *Nesokia indica* (extant form). It differs from *Bandicota* in that it lacks an anterostyle on M1 and a labial anterocone on M2. An attempt has been made to distinguish species of *Nesokia* and *Bandicota* based on size variation (*fig.* 8).

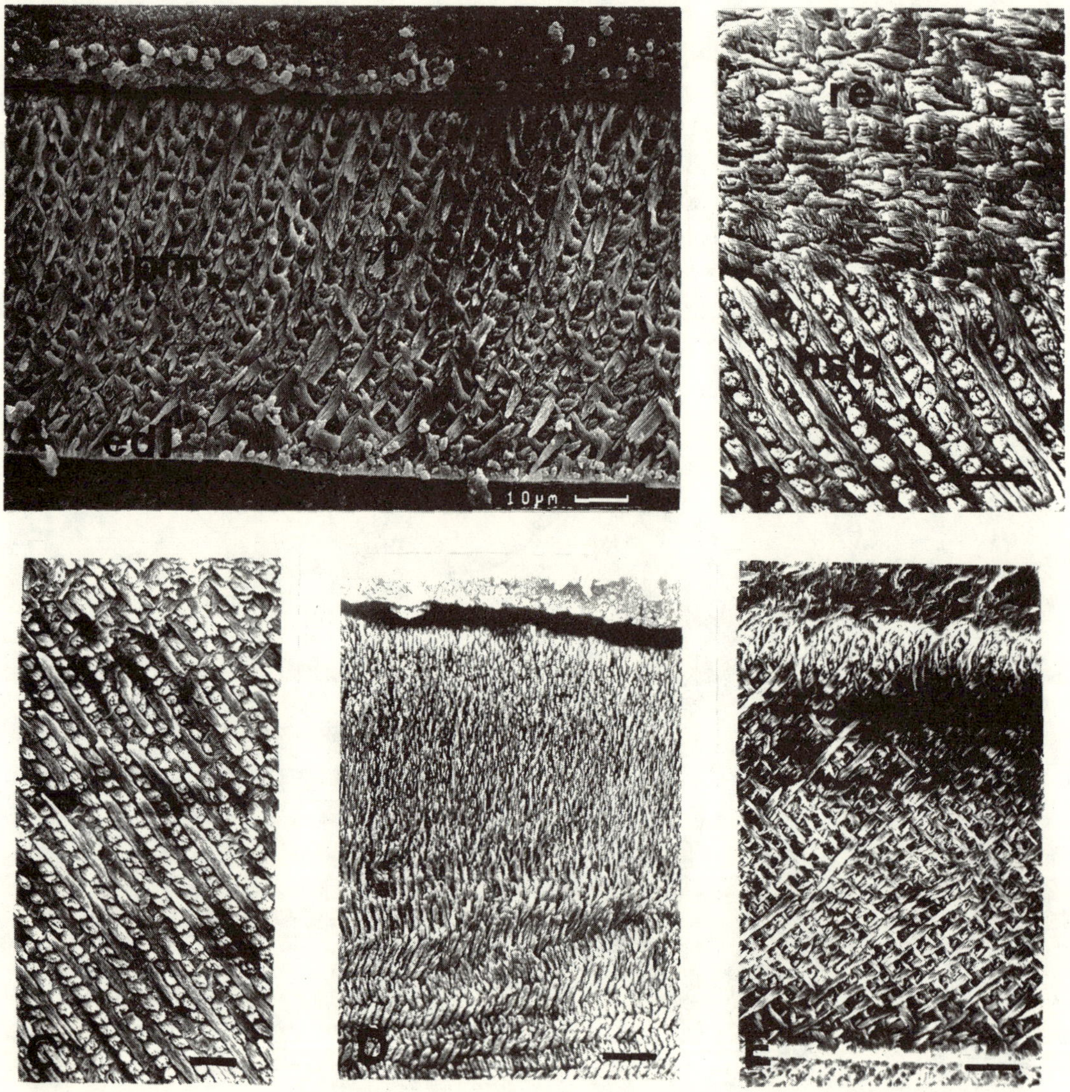

fig. 7 **A.** Longitudinal section of upper incisor of *Tatera* cf. *T. indica;* B-E, *Bandicota* cf. *B. bengalensis;* **B.** Upper portion of longitudinal section of upper incisor; **C.** lower part in detail; **D.** upper portion of the transverse section of upper incisor; **E.** longitudinal section of the lower incisor. Bars represent 10 mm.

Strong decussating prisms of alternating Hunter-Schreger Bands provide resistance against crack-producing forces, and thick outer radial enamel resists wearing due to cutting action. Prisms are strongest along their c-axis and in the radial enamel they are arranged parallel to each other, directed towards the cutting edge.

Class : Mammalia

Order : Rodentia

Family : Gerbillidae

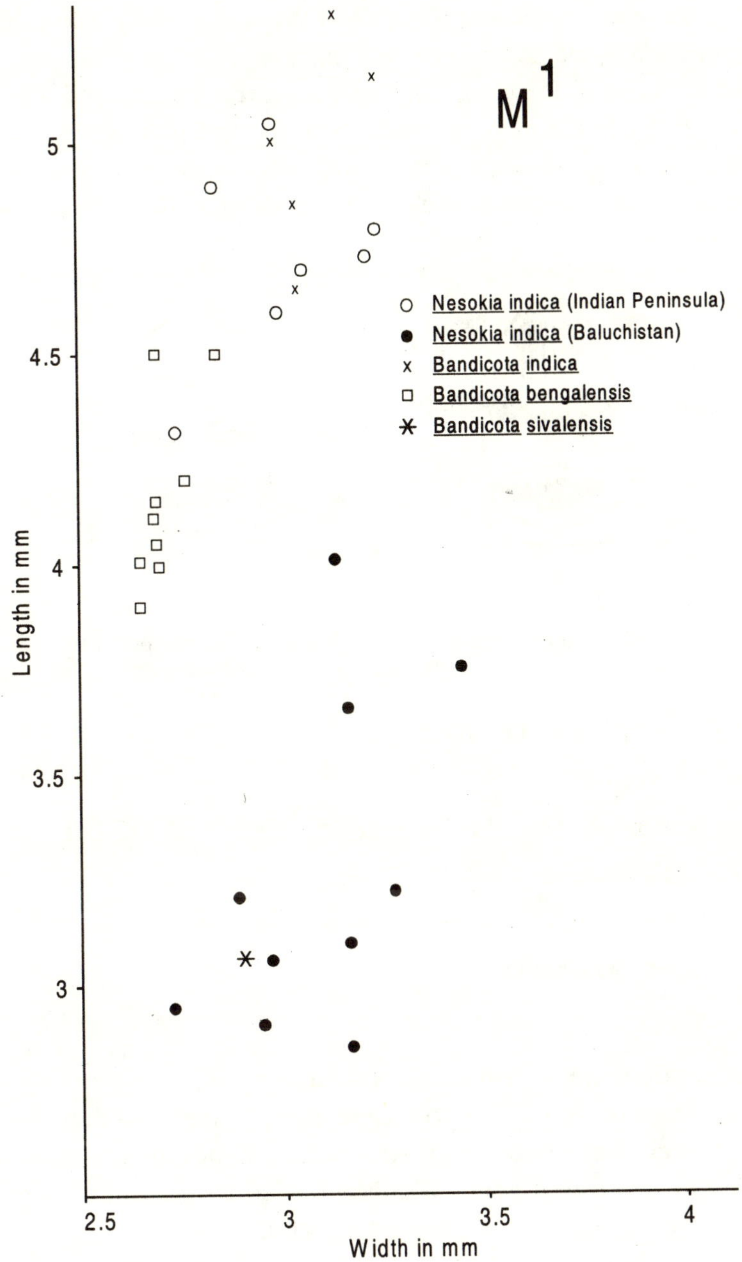

fig. 8. Scatter diagram showing size variation within *Nesokia* and *Bandicota* extant specimens, *Bandicota indica, Bandicota bengalensis,* and *Nesokia indica* and fossil specimen, *Bandicota sivalensis* (from Patnaik 1991).

Gerbillids are widespread in the African savannah region. In Asia, they inhabit a large region particularly in the arid zones. They are quite common in Indian deserts. The few finds of fossil gerbillids from the Indian subcontinent include *Protatera kabulense* from the Pliocene sediments of Afghanistan (Sen, 1983). *Tatera pinjoricus, Tatera abucnabia* cf. *A. kabulense* from the Late Pliocene Śiwālik sediments of India (Patnaik, 1997). The Afghan *Protatera* has recently been included in the genus *Abudhabia* (Bruijn and Whybrow, 1994). Barry and Flynn (1990) have mentioned the record of *Protatera* from Late Miocene Śiwālik sediments of Pakistan. *Protatera, Abudhabia* are extinct forms and are believed to have given rise to the modern gerbillids of the Indian subcontinent. Molars of gerbillids show great intraspecific variation, hence it is very difficult to taxonomically classify gerbillids solely on the basis of a few molars (Wesselmen, 1984 ; Denys 1989).

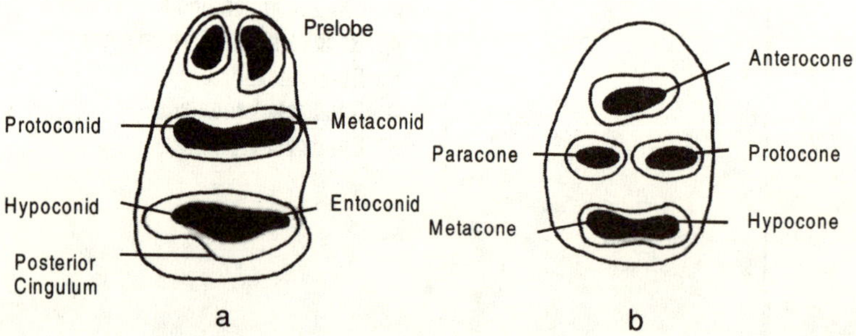

fig. 9. Gerbillid dental terminology, a, M and b. M1 (after Tong, 1986).

Genus - *Tatera* LATASTE, 1882

Tatera cf. *T. indica*

Referred material : One maxilla with M1, M2, M3 (PLDC-G-1)

Locality : Devakachār

Stratigraphic range : Devakachār Formation

Age : Upper Pleistocene

DESCRIPTION

M1 comprises three transverse cheverons (ordaminae). The two cusps of the prelobe are fused together. The protocone and paracone are almost equal in size and strongly connected. The metacone and hypocone are of almost the same size and are transversely aligned. A distinct posterior cingulum is attached to the hypocone by a strong connection.

M2 consists of two transverse cheverons. The protocone and paracone are very strongly joined (actually fused). The hypocone and metacone are also strongly connected. The posterior cingulum is distinct and attached to the hypocone.

M3 is sub-triangular in occlusal outline and comprises protocone and paracone (first cheveron) and a second hypocone and metacone (*fig.* 6).

The enamel microstructural study reveals presence of gently inclined Hunter-Schreger Bands occupying almost the whole of the enamel (*fig.* 7 A). The prisms of alternating bands decussate and form a plywood kind of pattern. The crystallites of inter-prismatic matrix run perpendicular to the prisms.

COMPARISONS AND DISCUSSION

Tatera indica (extant Indian gerbillid), the only species to have been included in the sub-genus *Tatera,* exhibits a permanent longitudinal connection between anterocone and the second lobe on M1. Fossil *Tatera* is known from the Late Pliocene sediments of India (Patnaik, 1997), Omo Shungura Formation (Wesselman, 1984), and the Hadar Formation of Ethiopia (Sabatier, 1982). Pleistocene deposits of Tanzania (Olduvai Bed 1) have yielded *Tatera gentryi* (Denys, 1989). The present *Tatera* cf. *T. indica* most closely resembles *Tatera pinjoricus.* It differs from *Tatera* sp. indet. (Wesselman, 1984) in having a relatively larger M1 with strongly fused cusps on the prelobe and a prominent posterior cingulum. M1 again has a more laminated anterior cheveron and consists of a posterior cingulum. On M1 the posterior cheveron is relatively larger in size due to the fusion of the hypocone and the metacone. As compared with *Tatera* sp. of the Hadar Formation (Sabatier, 1982), the *Tatera* has a more prominent posterior cingulum on M1 and M2 . The other features are, however, quite similar. *Tatera gentryi* is a very well studied form (Denys, 1989). Its upper molars resemble those of the living *Tatera* except that the former lacks the posterior cingulum, and the posterior cheveron on M3 is reduced to one cusp.

The enamel microstructural pattern suggests a compact and stable system which provides strength to the incisor against tensile stress developed due to repeated cutting of food material.

Gerbillus DESMAREST, 1804,

Gerbillus sp.

Referred material: Left maxillae with M2 (PLDC-G-2)

Locality : Devakachār

Stratigraphic range : Devakachār Formation

Age : Upper Pleistocene

DESCRIPTION

M3 has a squarish occlusal outline and comprises two well laminated cheverons. The protocone and the metacone are joined strongly. The hypocone and the metacone have a strong transverse connection. The two cheverons are connected longitudinally (*fig.* 6).

COMPARISONS AND DISCUSSION

Gerbillus is characterized by the presence of a permanent longitudinal link between the different lobes. M2 in the present specimen differs from that of *Gerbillus indus* (extant taxon) by having almost vertical cheverons, and the cusps are alternate to a lesser degree. M2 of *Gerbillus* sp. reported from Olduvai Bed 1 (Lower Pleistocene), Tanzania (Denya, 1989) differs from that of the present specimen in having a posterior cingulum, an anterior mure and lacks a distinct longitudinal connection.

Class : Mammalia

Order : Insectivora

Family : Soricidae

Today, shrews have a worldwide distribution. They are found throughout the Old and New World and are absent only in the Polar regions, the West Indies, certain islands in the Pacific, Australia, Tasmania, New Zealand and almost all of South America (Herter, 1972). Indian soricids belong to two sub-families, *Crocidurinae* and *Soricinae*. Shrews belonging to the sub-family *Crocidurinae* are distinguished by their unpigmented teeth, specific characters of the lower fourth premolar and the mandibular condyle (Repenning, 1967). On the other hand, shrews belonging to the sub-family *Soricinae* have pigmented teeth, a lingual condylar emargination of the mandible which is open, and a narrow interarticular area. The mental foramen lies beneath M1, M2 with prominent entoconid crest and upper incisor fissident (Repenning, 1967).

Fossil shrews of India are known from the Middle Śiwāliks (*Śiwālikosorex prasadi* Sahni and Khare, 1976); the Karewas (*Indosuncus bhatiai* Sahni and Kotlia, 1985) and *Episorictilus_repenningi* (Kotlia, 1991); and the Upper Śiwāliks (*Crocidura* sp., *Suncus*, and *Soricidae* indet. (Patnaik, in press).

Soricidae Indet

Referred material: A fragmentary right mandible (PLDC-11)

Locality: Devakachār

Stratigraphic range: Devakachār Formation

Age : Upper Pleistocene

DESCRIPTION

The soricid mandible lacks teeth and its condyle is broken. However, all the roots are preserved (*fig.* 10 a).

COMPARISONS AND DISCUSSION

The present mandible most closely resembles that of a Crocidura in size and shape. Due to the incompleteness of the specimen, it is not possible to classify it.

Class : Reptilia

Order : Crocodylia

Family : Crocodylidae

Present-day fresh water crocodiles of India include *Crocodylus palustris* and *Gavialis gangeticus*. Śiwālik sediments have yielded numerous specimens of *Crocodylus* (see Patnaik and Schleich, 1993) for review. *Crocodylus* is also known from Quaternary sediments of the Narmadā and the Godāvarī valleys and the Kurnool Caves (Prasad, 1982; Badam, 1984).

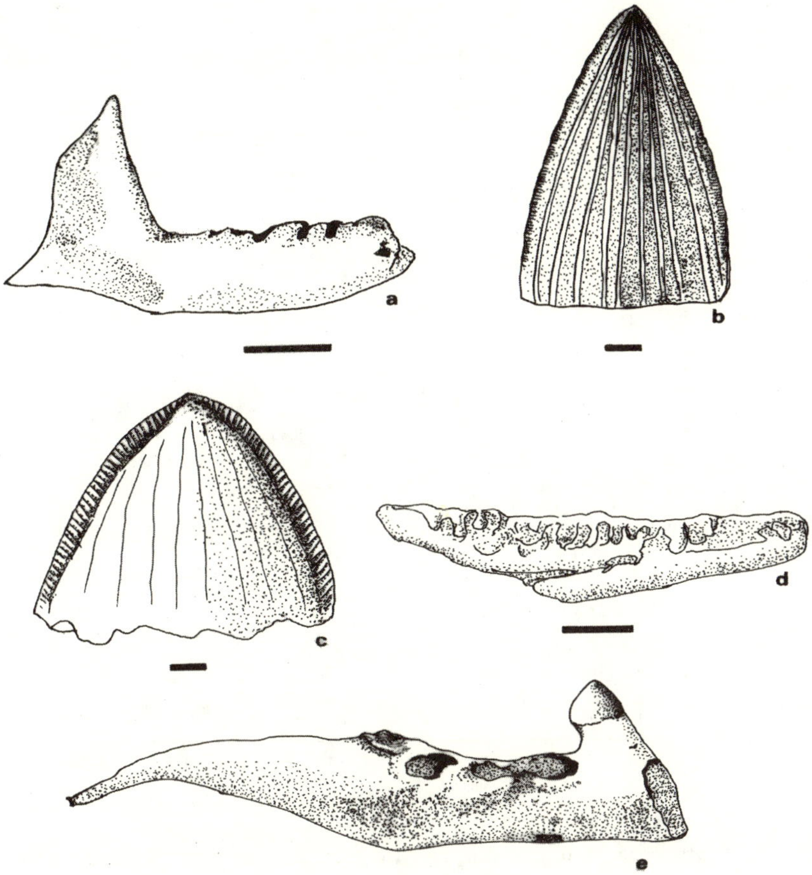

fig. 10. Soricidae indet. **a**. labial view of the mandible (PLDC-1-1). *Crocodylus* cf. *C. palustris*. **b**. anterior tooth (PLDC-R-1) and **c**. posterior tooth (PLDC-R-2). Anura indet. **d**. linguo-occlusal view of the maxilla (PLDC-A-1). *Cyprinus* cf. *c. Carpio cummunis*. **e**. pharangeal arch with one tooth (PLDC- P-1). Bars represent 1 mm.

Crocodylus **Laurenti 1768**

Crocodylus cf. *C. palustris*

Referred material : Two isolated teeth (PLDC-R-1 & 2)

Locality : Devakachār

Stratigraphic range : Devakachār Formation

Age : Upper Pleistocene

DESCRIPTION

One tooth (no. PLDC-R-1 *fig*. 10 b) shows vertical striation on the enamel surface. A faint keel can be seen on the two edges extending up to the sharp apex. The tooth is laterally compressed. The second tooth (no. PLDC-R-2: *fig*. 10 c) is blunt and the vertical striations are very faint. The lateral keel is very distinct and ornamented.

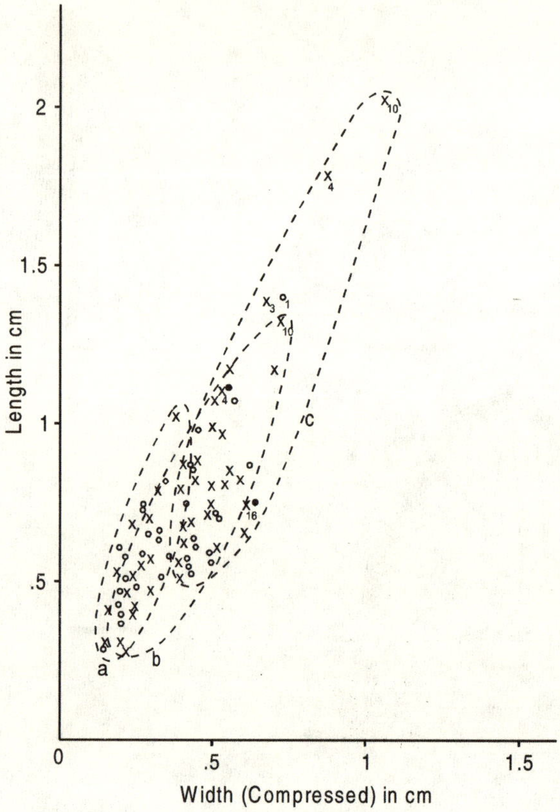

fig. 11. Scatter diagram showing teeth dimensions of *Crocodylus palustris* (o-maxillary and x-mandibular teeth) and *Crocodylus* cf. *C. palustris* (o-mandibular teeth). **a.** ZSM 3411912; **b.** ZSM 51711911 and **c.** ZSM 23 110 (modified from Patnaik and Schleich 1993).

COMPARISONS AND DISCUSSION

Both the teeth (PLDC-R-1, and 2) closely resemble the anterior (no. 4) and posterior (no. 16) mandibular teeth of *Crocodylus palustris* respectively. The dimensions of these teeth fall very much within the range of those of the present *Crocodylus palustris* (*fig*. 11).

Class	:	Osteichthyes
Order	:	Cypriniformes
Family	:	Cyprinidae

Cyprinids are quite common in fresh water ponds, lakes and stream systems. Fossil cyprinids in India are known from the Karewas of Kashmir (Kotlia, 1984) and the Śiwāliks (Raghavan, 1988; Patnaik, 1991).

Cyprinus* cf. *C. carpio communis

Referred material : Partly broken pharangeal arch with a single tooth (PLDC-P-1)

Locality : Devakachār

Stratigraphic range : Devakachār Formation

Age : Upper Pleistocene

DESCRIPTION

The right pharangeal arch consists of a single medium-sized tooth. Three alveoli are quite distinct. The anterior limb (ceratobranchial) is broken. The tooth shows a semicircular to conical outline (*fig.* 10 e).

COMPARISONS AND DISCUSSION

The present pharangeal arch closely resembles that of *Cyprinus carpio communis* Type 2 teeth described by Kotlia (1984) from the Karewas of Kashmir.

Class : Amphibia

Order : Anura

Family : Anura indet

Referred material : Single maxilla fragment (PLDC-A-1)

Locality : Devakachār

Stratigraphic range : Devakachār Formation

Age : Upper Pleistocene

DESCRIPTION

The maxilla fragment shows a typical dental gutter on the lingual side. The labial side is non-sculptured. The fragment is highly abraided and no tooth is preserved (*fig.* 10 d).

COMPARISONS AND DISCUSSION

The jaw fragment resembels maxillae of both fossil and extant frogs.

Palaeobiogeography of Narmadā Rodents

Murids (rats and mice), which are found in the Pleistocene sediments of Narmadā today, originated in the Śiwāliks of the Indian subcontinent some 13.75 myr ago. The oldest known Śiwālik taxa *Antemus chinjiensis* arose from a cricetid *Potwarmus primitivus* of Chinji Formation at Potwar Plateau. Though man is responsible for their present-day distribution, *Progonomys* (which evolved from *Antemus* in the Potwar Plateau and is considered to be the descendant of around 500 murid species found today) began its journey some 11 myr ago from the Indian subcontinent to Europe, Africa and China (Jacobs and Downs, 1994). Till the uppermost Miocene, Śiwālik murids were not very diverse and were represented only by three identifiable taxa *Mus*, *Karnimata*, and *Parapelomys*. This was the time when *Mus* (sub-genus *Nannomys*) or an ancestor of the genus *Mus* migrated westwards to appear in Afghanistan and later in Kenya, Ethiopia, and Morocco. The Late Miocene gerbillid *Abudhabia* spread eastwards from Abu Dhabi, UAE (Bruijn and Whybrow,

1994) to Kabul, Afghanistan (Lowermost Pliocene), Moginand, Himachal Pradesh (India) around 4 myrs ago. *Dilatomys* the first hypsodont murid) of Early Pliocene from Kabul also used the same route to reach Moginand, where *Golunda* and *Cremnomys* made their first appearance.

Within India, the Narmadā rodents have common elements in the Śiwāliks, Karewa Basin, Bhīmtāl Lake, and Kurnool Caves. They resemble also those found outside India (Potwar Plateau, Pakistan; Pul-Charkhi Basin, Kabul, Afghanistan; Abu Dhabi, United Arab Emirates; Omo Shungura and Hadar, Ethiopia; Lake Turkana, Kenya and Casablanca, Morocco). Around Late Pliocene, *Golunda* and *Mus* migrated to the Karewa Basin of Kashmir. During the Upper Pleistocene, murids had a wider distribution; *Mus*, *Millardia*, *Bandicota* and *Tatera* (a gerbillid) spread southwards to the Narmadā; *Mus* and *Golunda* to Kurnool Caves; *Mus* to Bhīmtāl region, Uttar Pradesh; Nesokia spread westwards to Palaeolithic deposits of Sudan.

Based on the closeness of fossil murids from Śiwāliks, Karewas of Kashmir (around 2.4 myrs), Narmadā Valley of Central India (around 35,000 years), Kurnool Cave deposits of southern India (around 40,000 years) and Bhīmtāl Lake sediments (around 30,000 years) to their extant counterparts, it can be

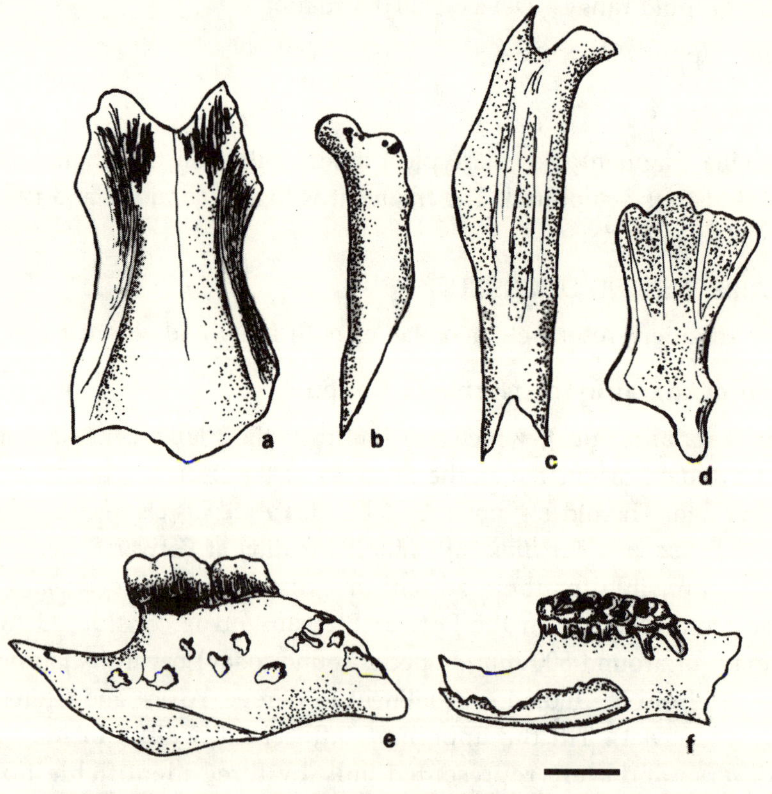

fig. 12. **a.** Skull cap of *Bandicota* cf. *B. bengalensis* (PLDC-M-10); **b** & **c.** Femora fragments of rodents; **d.** Scapula of rodent; **e.** and **f.** lower jaws of *Millardia* cf. *M. meltada* Bar represents 2 mm.

safely concluded that the Pliocene and Pleistocene murids recovered so far were endemic to the Indian subcontinent.

Taphonomic Considerations

Mellet (1974) proposed that almost all microvertebrate accumulations are essentially biogenic (deposited by Carnivores) to begin with and may eventually be transported and re-deposited by fluvial processes. He used the term "coprocoenosis" for this process, which results in the fracture and abrasion of bones, earlier suggested as being seen only on bones accumulated by fluvial processes (Dodson 1974; Wolf 1981).

To ascertain whether a micorvertebrate accumulation is of fluvial or biogenic origin a comprehensive experimental analysis was undertaken by Korth (1979). According to him, in accumulations resulting from coprocoenosis one should expect a high percentage (greater than 60 per cent) representation of the bones of the animal; the majority of limb bones should be complete; skulls should occur commonly and frequently and should be complete, and there should be no evidence of mechanical abrasion. In the case of accumulations due to fluvial processes, the sediments enclosing the fossils should generally show primary sedimentary structures. The model size of sand grains in the deposit should be predictable from the size and hydraulic characteristics of the skeletal elements present; the bones should show varying degrees of stream abrasion. The percentage representation of bones of all the fauna should be low, and a particular element or elements may be markedly abundant for each individual taxon.

Taphonomic interpretations of the present microvertebrate assemblage are based on the above mentioned facts and comparisons with various Pleistocene microvertebrate accumulations.

The only skull fragment in the present collection (*fig*. 12 a) resembles that extracted from an owl pellet rather than to that fluvially abraided (Korth, 1979: *fig*. 5 a; Dodson and Waxler, 1979: *fig*. 3 b). The angle, coronoid and condyle of most of the mandibles have been worn down (*fig*. 12 e-f). There is a thinning of bones, and the sockets of the molars and pulp cavity for the incisors commonly show holes at the base and on the ventral surface of the jaw around the incisors (*fig*. 12 c and f). In the present assemblage, a high percentage of limb bones are present, some of which show angular breakages (*fig*. 12 b-c), very similar to those shown by bones extracted from coyote (*Canis*) faeces (Korth 1979). Bones like scapulae, humerii and pelvis show frequent breakages.

Besides the above mentioned features, there are certain observations which cannot be ignored. There is a high proportional representation of incisors, tibia, femur and humerus bones (Table 2). The distribution of skeletal elements shows almost an entire range of micromammalian skeletal parts being represented here together with a few elements of fish, crocodile, turtle and from remains, well rounded pebbles, concretions and haphazardly

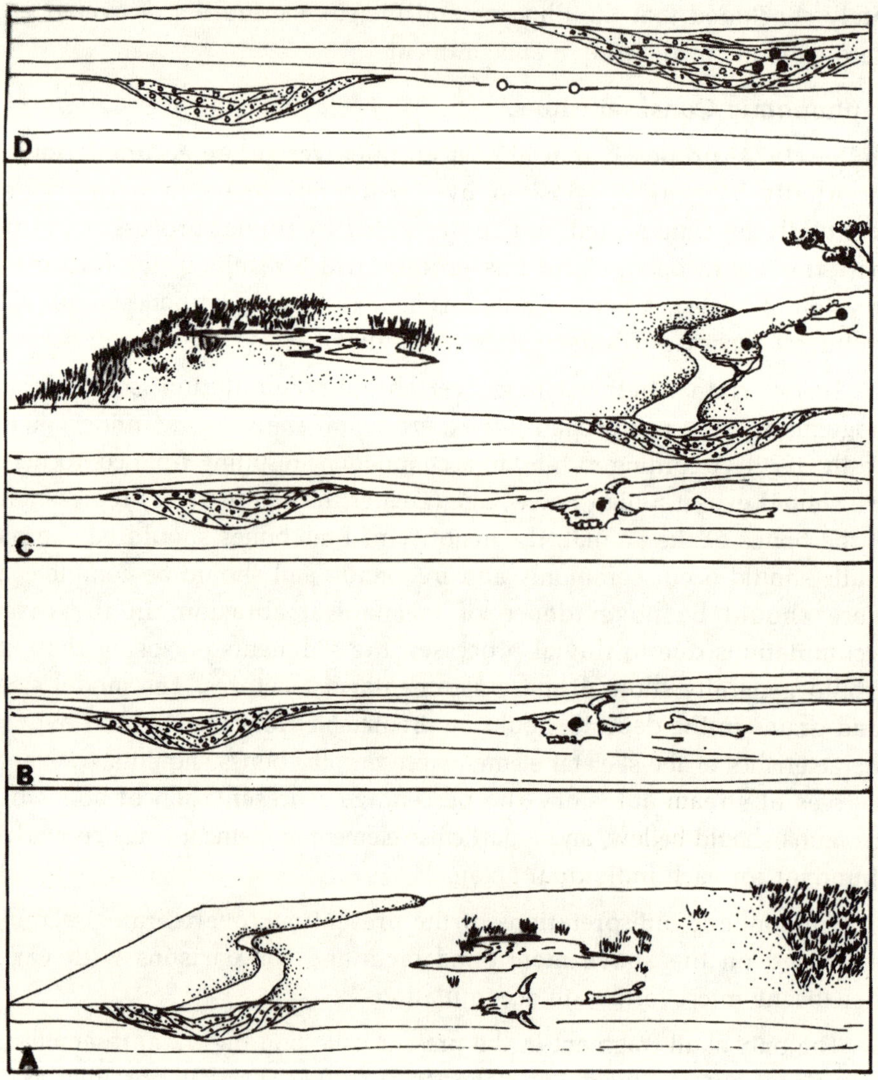

fig. 13. Hypothetical reconstruction showing the various stages of fossil accumulation and sedimentation. **A.** accumulation of large mammal remains in a flood-plain area. **B.** flood buries the large mammal remains. **C.** lateral migration of the stream channel and accumulation of small mammal remains. **D.** flood facilitates small scale transportation and deposition of small vertebrate remains, large mammal remains and small vertebrate remains.

oriented molluscan shells. This microvertebrate horizon shows typical characteristics of a sandy pebbly or a conglomeratic channel deposit (Khan, 1984). Some bones like scapulae, humerii, femora and pelves show abrasion on their ends and processes which can only be seen on bones derived from fluvial deposits (Korth, 1979).

In the light of the observation made, it can be concluded here that the micromammals were primarily derived from pellets of carnivorous birds like owls, and mammals like coyote. Recently, a humerus of Scops Owl (?) has been recorverd from the same horizon (Patnaik and Sahni, 1994). Scops owls are widely distributed in Peninsular India and feed on small mammals and insects. This find may support the above-mentioned interpretations.

After the disintegration of the pellets, these skeletal remains were transported and finally deposited by fluvial processes responsible for the transportation and deposition of the pebbles, concretions and sandy matrix of the conglomeratic layer. At Devakachār, the dark brown mudstones (*fig.* 1) have yielded an almost complete skull of a bovine without the evidence of abrasion or transportation (Badam: personal communication). To explain the various stages of large vertebrate and microvertebrate accumulation at Devakachār, a hypothetical model has been suggested (*fig.* 13).

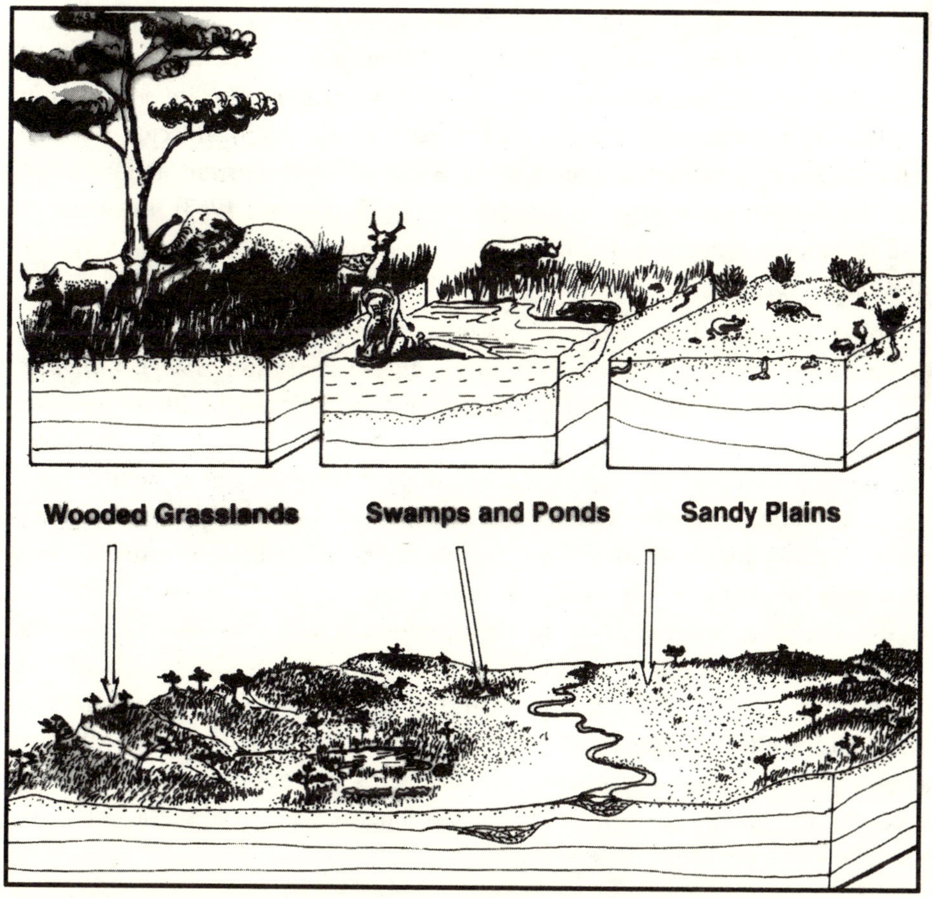

fig. 14. Hypothetical reconstruction of the various communities present and deposition of Devakachār Formation in the Narmadā Valley.

The Hathnorā community basically represents a pond/seasonal pool environment proximal to the channel area. There was not much transportation involved.

Palaeoecological Implications

Palaeoecological reconstructions based on micromammal assemblages are quite accurate as they are found confined to beds of small lateral extent, deposited in a short period of time. Micromammals are very sensitive to climate changes and usually have small home ranges, which makes them very useful climatic indicators. In the present work, an attempt has been made to integrate local palaeoecological interpretation based on microvertebrates with the regional perspective provided by megavertebrates. Sedimentological observations have also been used here to provide a better resolution of the ecological conditions prevailing during the deposition of the Devakachār Formation. Since most of the microvertebrates resemble extant forms, the palaeoecological implications are based on the principle of actualism.

At Hathnorā, the thick piles of clays and occasional cross-bedded pebbly layers represent predominant deposition under flood plain conditions. *Darwinula stevensi* and *Candona fabaeformis* show affinity for permanent still water bodies and muddy ponds and ditches with moderate vegetation (Staplin, 1963; Bhatia and Singh, 1971; Bhatia et al., 1985). *llyocypris bradyi* prefers temporary or permanent running water or some current action in the water bodies with rich vegetation (Holmes, 1937; Staplin, 1963). Presence of small streams and ponds is indicated by the mollusca, *hydrobiidae* and *pisididae* indet. and *cyprinid*.

Today *Bandicota bengalensis* occurs only in irrigated croplands where the sub-soil moisture is very high throughout the year (Prater, 1971). *Millardia meltada* prefers cultivated fields and sometimes heavy shrubs and rocks. According to Prakash (1975), *Gerbillus indus* occurs in a sandy habitat. *Tatera indica* prefers sandy plains and interdunal regions where it can dig extensively and develop deep burrow systems. In Africa, *Tatera* is found primarily in dry steppe country and sometimes in thickets along the edges of alluvial flats bordering the Kerio River in South Turkana (Coe, 1972; Hubert, 1978). In general, gerbillids are savannah rodents preferring arid and desertic regions. *Gerbillus pusilus* like other east African gerbills live in the drier parts of this region. They live in open sandy soil and alluvial areas which flood in wet season and dry out during summer. Shrews like *Crocidura* have wide ecological ranges extending from moist forest to scrub land and appear to be extremely successful in areas of ecological diversity as long as they have sufficient cover in thickets or vegetation (Wesselman, 1984).

Crocodiles, cyprinids and frogs indicate the presence of a well watered landscape and a warm climate.

Large vertebrate fauna reported so far by various authors suggest the presence of open grasslands and wooded grasslands interspersed with perennial rivers and swamps (Badam, 1984).

At the microvertebrate yielding locality (*fig.* 2), the sandy pebbly conglomeratic layer contains haphazardly oriented molluscan shells, indicative of the channel area of a stream. The overlying and underlying pedogenically modified clays containing concretions suggest their development in a floodplain area away from streams, ponds and swamps. In fact an attempt has been made here to illustrate the various communities present, major vegetation types and the depositional environment of the Devakachār Formation (*fig.* 14).

To sum up, the famous hominid bearing Pleistocene deposits of Narmadā Valley have yielded microfaunal assemblages from two important localities, namely, Hathnorā and Devakachār of Hoshangabad and Narsinghpur districts respectively of Madhya Pradesh. The microvertebrates recovered from Devakachār and described earlier have been updated in the light of new approaches and developments made over the last five years. Microfossils such as ostracodes, *Candona fabaeformis, Darwinula stevensi, llyocypris bradyi* and *Cypridopsis* sp.; molluscs belonging to *Hydrobiidae* indet., cf. *Gyraulus, Pisididae* indeterminate, discovered for the first time from the hominid bearing Hathnorā locality, indicate presence of permanent lakes/ponds with occasional influx of running water with current action and moderate to rich vegetation in the area. At Devakachār, palaeoecological reconstruction indicates the presence of sandy plains, grasslands with a high sub-soil moisture content and shallow stream systems. The micromammal remains at Devakachār were accumulated primarily by carnivorous birds and mammals and were later transported and deposited by fluvial processes along with those of associated microvertebrates. The Narmadā Valley microfauna is not endemic and has common elements in Karewas of Kashmir, Upper Śiwāliks and Kurnool Caves. Outside India, the generic counterparts of some of the Narmadā rodents are found in Plio-Pleistocene deposits of Pakistan, Afghanistan, Ethiopia, Kenya, Tanzania and Morocco.

Table 1

Mammalia	**Rodentia**	Muridae	*Millardia* cf. *M. meltada* *Bandicota* cf. *B. bengalensis*
		Gerbillidae	*Tatera* cf. *T. indica* *Gerbillus* sp.
	Insectivora	Soricidae	*Soricidae* indet.
Reptilia	**Crocodylia**	Crocodilidae	*Crocodylus* cf. *C. plustris*
Pices	**Cypriniformes**	Cyprinidae	*Cyprinus* cf. *C. carpio communis*
Amphibia	**Anura**	Anura indet.	

Table 2
Numbers and proportional percentage of small mammal skeletal elements

Element	Number	Percentage
Skull	01	0.19
Mandible	37	7.25
Maxilla	22	4.31
Incisor	223	43.72
Molar	04	0.78
Scapula	06	1.17
Humerus	44	8.62
Radius	30	5.88
Ulna	12	2.35
Vertebrae	07	1.37
Pelvis	08	1.56
Femur	50	9.80
Tibia	53	10.39
Calcaneum	01	0.19
Metapodials	12	2.35

Abbreviations

PLDC — M, Palaeontology Laboratory Deccan College, Murids; G — Gerbillids, I — Insectivores; R — Reptiles, P — Pices, A — Amphibians; ZSM, Zoological State Collection (Munich). VPL/RP, Vertebrate Palaeontology Laboratory/Rajeev Patnaik. p — Prisms; ipm — inter-prismatic matrix; edj — enamel-dentine junction; re — radial enamel; hsb — Hunter-Schreger Bands.

Acknowledgement

I am grateful to Prof. A. Sahni for valuable suggestions on the manuscript. I would also like to extend my gratitude to Dr G.L. Badam, who generously provided the microvertebrate material collected under his supervision for the present study. Thanks are due to Dr V. Sathe for his help both in the field and laboratory. I would like to thank Prof V. Fahlbush, Dr Thomas Bolliger and Dr Wenyu Wu (Munich) for fruitful discussions. Many Institutions provided their library, laboratory and musuem facilities for the present study. They include: CAS in Geology, Panjab University, Chandigarh; Deccan College, Pune; Bombay Museum of Natural History, Bombay; Zoological Survey of India Collection, Kolkatta; Insititute of Palaeontology and Historical Geology, and Zoological State Collection, Munich; Institute of Palaeontology and Koenigs Museum, Bonn. Financial support from the CSIR (New Delhi), DAAD (Bonn) and DST (New Delhi) at various stages of the research is thankfully acknowledged.

References

Agrawal, D.P., B.S. Kotia & S. Kusumgar, 1988, "Chronology and Significance of the Narmadā Formations," *Proceedings of Indian National Science Academy* 54(3); pp. 418-24.

Badam G.L., 1984, "Pleistocene Faunal Succession of India," in *The Evolution of East Asian Environments* (R.O.Whyte, ed.), pp. 746-75, Hong Kong : Centre of Asian Studies, University of Hong Kong.

Barry, J.C. and L.J. Flynn, 1990, "Key biostratigraphic events in the Śiwālik sequence," in *European Neogene Mammal Biochronology* (Lindsay, E.H., V. Fahlbusch & P. Mein, eds.), New York: Plenum Press.

Bhatia, S.B., 1969, "A New pulmonate (Pupillid) Gastropod from the Upper Karewas of Kashmir," *Bulletin of Geological Society of India* 6(4) : pp. 117-19.

Bhatia, S.B. & D. Singh, 1971, "Ecology and distribution of some recent ostracodes of the Vale of Kashmir, India," *Micropalaeontology,* 17(2): pp. 214-20.

Bhatia, S.B., R.S. Batra & B.S. Kotlia, 1985, "Ostracodes and Palaeoecology of the Karewa Formation of Hirpur and Romushi Sections, Kashmir, India," *Current Trends in Geology* VI : pp. 45-52.

Bruijn, H.D. and P.J. Whybrow, 1994, "A Late Miocene rodent fauna from the Baynunah Formation, Emirates of Abu Dhabi, United Arab Emirates," *Proc. Kon. Ned. Akad. V. Wetensch.* 97 (4) : pp. 407-22.

Coe, M, 1972, "The South Turkana Expedition, Scientific Papers IX; Ecological Studies of Small Mammals of South Turkana," *Journal of Geography* 138: pp. 316-38.

Denys C., 1989, "Two New gerbillids (Rodentia, Mammalia) form Olduvai Bed I (Pleistocene, Tanzania)," *Neues Jahrbuchfur Geologie and Pcilaoiitologie* 178 (2): 243-65.

Dodson P., 1974, "The Significance of Small Bones in Palaeoecological Interpretation, University of Wyoming," *Contribution of Geology* 12: pp. 15-19.

Dodson, P. and D. Waxler, 1979, "Taphonomic investigations of Owl Pellets," *Palaeobiology* 5 (3): pp. 275-84.

Herter, K., 1972, in *Grzinieks Animal Life Encyclopaedia* 10 : pp. 176-257, New York: Van Nostrand Reinhold Company.

Holmes, P.F., 1937, "Ostracoda from Lake Ohrid, *Archir,*" *Hydrobiology* 31: pp. 484-500.

Hubert, B., 1978, "Modern Rodent Fauna of the lower Omo Valley, Ethiopia," *Bulletin Carnegie Museum of Natural History* 6: pp. 109-12.

Jacobs, L.L., 1978, "Fossil rodents (Rhizomyidae and Muridae) from Neogerne Śiwālik Deposits, Pakistan," Museum Arizona Press, Flagstaff, Bulletin Series 52: pp. 1-103.

Joshi, R.V. & A. Kshirsagar, 1986, *Chemical Studies of Archaeological Bones from India: Flourine and Fossilisation Process,* Pune: Deccan College.

Khan, A.A., 1984, "Geological and Geomorphological studies around Tapti-Vagher Confluence, District Jalgaon, Maharashtra," *Geological Survey of India Review* 113: pp. 99-109.

Khan, A.A. and A. Sonakia, 1992, "Quaternary Deposits of Narmadā with special reference to the Hominid Fossils," *Journal Geological Society of India* 39: pp. 147-54.

Korth, W.W., 1979, "Taphonomy of Microvertebrate Fossil Assemblages," *Annals Carnegie Museum* 48: pp. 235-85.

Kotlia, B.S., 1984, "Vertebrate Palaeontology and Palaeoecology of the Karewa Group, Kashmir, a Biostratigraphical Study," Ph.D. Thesis. Chandigarh: Panjab University.

———, 1991, "Pliocene Soricidae (Insectivora, mammalia) from Kashmir Intemontane Basin, Northwestern Himalayas," *Journal Geological Society of India* 38 (3): pp. 253-357.

———, 1992, "Pliocene Murids (Rodenlia, Mammalia) from Kashmir Intemontane Basin, Northwestern India," *Neues Jahrbuchfur Geologie un. Palaeontologie* 184 (3): pp. 339-57.

Lumeley, M.A. & A. Sonakia, 1985, "First discovery of Homo erectus on the Indian subcontinent, at Hathnorā, in the Middle Valley of the Narmadā River," *L'Anthropologie (Paris)* 89: pp. 13-61.

Mellet, J.S., 1974, "Scatalogical Origin of Microvertebrate Fossil Accumulations," *Science* 185: pp. 349-50.

Missone, X., 1969, "African and Indo-Australian Muridae, Evolutionary Trends, Muse Royal de l'Afrique Centrale-Tervuren, Beigiqtie," *Annales Sciences Zoologiques* 8(172): pp. 1-219.

Patnaik, R., 1993, *Micropalaeontology, Biostratigraphy and Palaeo-environmental Analysis of the Śiwālik Sequences in the Saketi-Nahan Area* (H.P), Ph.D. Thesis, Chandigarh: Panjab University.

——, (in press), Late Pliocene micromammals from Tatrot Formation (Upper Śiwāliks) Exposed near Village Saketi, Himachal Pradesh, India, *Palaeontographica.*

——, H.H. Schleich, 1993, "Fossil Crocodile Remains from Upper Śiwāliks, India," *Mitteilungen der Bayerischen Staatssamnilungfut Palaeontologie and Historiche Geologie* 33: pp. 91-117.

——, G.L. Badam & V. Sathe, 1995, "Discovery of Microvertebrates frm the Pleistocene Deposits of the Central Narmadā Valley, India," *Current Science* 68 (8): pp. 828-30.

——, A., Sahni, 1994, "Record of a Bird Humerus from Upper Pleistocene Narmadā Valley Sediments," *Palaeontological Society of India* 39: pp. 77-79.

——, 1995, "Narmadā Valley Microvertebrates: Systematics, Taphonomy and Palaeoecology," *Man and Environment* XX (2): pp. 75-90.

——, 1997, "New Murids and gerbillids (Rodentia, Mammalia) frm Pliocene Śiwālik sediments of India," *Palaeovertebrata* 26 (1-4): pp. 129-65.

Prakash, I., 1975, "The Population Ecology of Rodents of Rajasthan Desert," in *Rodents in Desert Environments* (Prakash, I. & P.K. Ghosh, eds.), pp. 75-114, The Hague: Dr. W. Junk, B.V. Publishers.

Prasad, K.N., 1982, "Studies on Neogene Quaternary Boundary in India and Adjacent Countries, Proceedings of Koninklijke Nederlandse Akaden 7,i c van *Wetensehappen* 85 (4): pp. 455-68.

Prater, S., 1971, *The Book of India Animals*, Bombay: Bombay Natural History Society.

Raghavan, P. 1988, "Palaeo-environment of the Pinjor Formation Based on Microfossil Assemblages and Enamel Ultrastructure of Pinjor and Other Śiwālik Primates," Ph.D. Thesis, Chandigarh: Panjab University.

Repenning, C.A., 1967, "Subfamilies and Genera of the Soricidae," *United States Geological Survey Professional Paper* 565: pp. 1-74.

Sabatier, M. 1982, Les Rongeurs du site Pliocene a Hominidi de Hadar (Ethiopia), *Palaeovertebrate,* 12 (1): pp. 1-56.

Sahni, A. & S. K. Khare, 1976, "Śiwālik Insectivora," *Journal Geological Society of India* 17: pp. 114-16.

Sahni, A. & B.S. Kotlia, 1985, "Karewa Microvertebrates: Biostratigraphical and Palaeoecological implications," in *Climate and Geology of Kashmir, The last 4 Million Years* (D.P., Agrawal, S. Kusumgar and R.V. Krishnamurthy, eds.), pp. 29-43, New Delhi: Today and Tomorrow's Publishers and Printers.

Sen, S. 1983, "Rongeurs et Lagomorphes du Gisement Pliocene de Pul-e Charkhi, Bassin de Kabul, Afghanistan," *Bulletin Museum Nationale Historie Naturelle de Paris* 5(1): pp. 33-174.

Sonakia, A., 1984, "The Skull Cap of Early Man and Associated Mammalian Fauna from Narmadā Valley Alluvium, Hoshangabad Area, Madhya Pradesh, India, Records," Geological Survey of India 113: pp. 159-72.

Sonakia, A. & S. Biswas, 1998, "Antiquity of the Narmadā Homo erectus, the early man of India," *Current Science* 75 (4): pp. 391-93.

Staplin, F.L., 1963, "Pleistocene Ostracoda of Illinois. Part-II Subfamilies Cyclocyprinae, Cyridopsinae," Stratigraphic ranges and assemblage patterns, *Journal of Palaeontology* 37 (6): pp. 1164-203.

Theobald, W., 1860, "On the Tertiary and alluvial deposits of the central portion of the Nerbudda valley," *Memoirs of the Geological Survey of India* 2: pp. 279-98.

Tong, H., 1986, "Origine et Evolution des Gerbillidae (*Mammalia, Rodentia) en Afrique du Nord.,*" Thesis Doctorate, Paris: University Paris.

Wesselman, H.B., 1984, "The Omo Micromammals. Systematics and Palaeoecology of Early Man Sites from Ethiopia," *Contribution to Vertebrate Evolution* 7: pp. 1-219.

Wolff, R.G., 1981, "Review of : Fossils in the Making," *Vertebrate Taphonomy and Palaeocology, in Paleobiology* (Behrensmeryer, A.K. & A.P. Hill, eds.), pp. 284-87, Chicago: University of Chicago Press.

Appendix II
Molluscan Shells

Arati Deshpande Mukherjee

Introduction

THE Central Narmadā region has been a subject of several geological, palaeontological and archaeological investigations since early 19th century. Several attempts were made to reconstruct the Pleistocene faunal history and palaeo-environment of the region. Discovery of a large number of vertebrate fossils and the recent findings of micromammals have yielded interesting information on the animal diversity and their ecosystems in the area.

Archaeological investigations were carried out to find the traces of hominid remains in the region. Today, there is sufficient evidence to confirm their presence in the Central Narmadā Valley in the form of stone artefacts belonging to Lower to Upper Palaeolithic periods, besides the discovery of the first hominid skull from late Acheulian cultural levels.

Besides vertebrate fossils, fresh water molluscan shells were recovered from this area. No specific studies of such shells have been carried out so far, except for their use in dating. However, molluscan shells from archaeological sites are known to yield a great deal of information regarding past subsistence (Meighan, 1969), palaeo-environment (Koike, 1977; Shackleton, 1970) and chronology (Evans, 1969). Fresh water and terrestrial molluscan shells in particular have been used in Europe and America for dating Quaternary deposits and for environmental reconstruction (Sparks, 1961; Burleigh and Kerney, 1982). With this in mind, shells from the Central Narmadā Valley were studied to help in reconstructing the Pleistocene environment in this particular region.

Region

The river Narmadā and its various tributaries like Sher, Umer, Varurewa, etc. have yielded rich fossiliferous sites between Hoshangabad and Narasinghpur in Madhya Pradesh. In this report, shells have been studied from the region of Devakachār in Dist. Narasinghpur. Devakachār (23° 23' N, 79° 07' E) is situated 10 km north-north-east of Narasinghpur on the right bank of the tributary Sher, slightly away from its confluence with another tributary, Umer. Both the tributaries flow towards north to join the river Narmadā. The entire area, both upstream and downstream of Devakachār to the confluence of various meandering streams like Sher, Umer, Varurewa, etc. has yielded fossils of *Elephas hysudricus, Elephas namadicus, Hexaprotodon* spp., *Equus namadicus, Crocodylus palaeindicus* and *Trionyx* sp. Fossil collection was made from excavations, river banks and gullies as well as from the adjoining areas of the tributaries.

Shells were collected as a part of the project titled, "Taphonomy and Palaeoecology of the Fossil Vertebrates from the Central Narmadā Valley. Dist. Narsinghpur, Madhya Pradesh, India." This was carried out jointly by the team of palaeontologists of the Department of Archaeology, Deccan College, Pune and the Indira Gandhi Rashtriya Manav Sangrahalaya, Bhopal in 1992-94. Shells were collected from three localities, Gurāriāghāṭ situated opposite to Devakachār, Guarighāṭ situated 50 m downstream of Gurāriāghāṭ and Talayyaghāṭ also located nearby, on the left bank of Sher-Umer.

The stratigraphy of the deposits of the tributary Sher at Devakachār is continuous with that of the fossiliferous alluvium of the Narmadā. Several researchers like Theobald (1860) and de Terra and Paterson (1939) have contributed immensely to determine the stratigraphy of the area.

The stratigraphy reveals that the lowermost unit in the exposed sections is the basal conglomerate, which has yielded Lower to Middle Palaeolithic tools. This is followed by the Red Brown concretionary silt which conformably overlies the gravels and encloses well preserved fossils belonging to the Older Alluvium of the Lower Narmadā Group. A deposit of sandy pebbly gravel rests unconformably on this reddish silt, forming the Upper Narmadā Group of the Older Alluvium, and has yielded fossils and Middle Palaeolithic tools. A partial skull of *Crocodylus palaeindicus* was recovered from this unit near Devakachār (Badam, 1990). Recently, micromammals comprising the rodent species *Golunda, Tatera, Rattus* and *Mus* were also recovered from this particular horizon at Gurāriāghāṭ (Patnaik et al., 1995). A radiocarbon date of 31750 (1770 ±1580) BP-TF 967 has been obtained for the fresh water shells from the sandy pebbly gravel at Ratikarar village (Agrawal and Kusumgar, 1974), the thickness of which varies from section to section.

This horizon is further overlain by yellow brown concretionary silt which is more sandy than the reddish silt. In this silt, belonging to the Younger Alluvium, a cranium of a gaur (*Bibos gaurus*) was discovered (Badam and Grigson, 1990). The top surface of the silt has weathered into the Black Cotton Soil. On the basis of fossil evidence and its dating and radiocarbon dating of shells, a greater part of the Narmadā Alluvium belongs to the Upper or Late Pleistocene (Badam, 1990). However, Middle Pleistocene deposits are now found exposed profusely in various sections.

Methodology

More than 500 samples of fresh water shells were randomly collected from three localities, Gurāriāghāṭ, Guarighāṭ and Talayyāghāṭ. Shells were hand-picked from the shelly pebbly gravel horizon exposed on the river banks. Some were collected from the small excavated trenches at Gurāriāghāṭ.

The shells were studied at the Archaeozoological laboratory of the Deccan College, Pune. A majority of shells show fairly good state of preservation. Some of the shells are complete with both valves intact and closed. Single valves are also represented. Gastropod shells are complete with a few specimens broken mostly at the apex and aperture area. The overall collection is mainly dominated by bivalves.

Not much mechanical cleaning was required since the shells were not encrusted in a hard matrix. Cleaning under running water was enough to remove the mud matrix sticking to them. Both Bivalves and Gastropods have lost their original colour, being dull chalky white to creamish, probably due to the surrounding sediments. In the bivalves, the nacreous layer has been exposed due to the removal of the periostracum layer.

Shells were sorted depending on species, rate of fragmentation, etc. and were then labelled and recorded. Identification was done by referring to Tonapi (1980), Subha Rao (1989) and to modern shell reference collection. However, there was difficulty in identifying and differentiating between two types of bivalve species because of their close similarity. Therefore, until further confirmation is made, these have been tentatively identified as *Lamellidens* sp. and *Parreysia* sp. Metrical analysis involved taking length and breadth measurements. However, length measurements could not be taken in most cases because the shells were broken lengthwise. The shells were also examined for human activity and other taphonomical marks.

Results

A total of 5 molluscan species have been identified, out of which three are bivalves and two are gastropods (Table 1), (*fig.* 1):

Table 1: Fresh water Molluscs from Central Narmadā Valley

Species	*Class*	*Family*
Lamellidens sp.	BV	Unionidae
Parreysia sp.	BV	Unionidae
Corbicula sp.	BV	Corbiculidae
Bellamya bengalensis	G	Viviparidae
Melania striatella tuberculata	G	Melaniidae

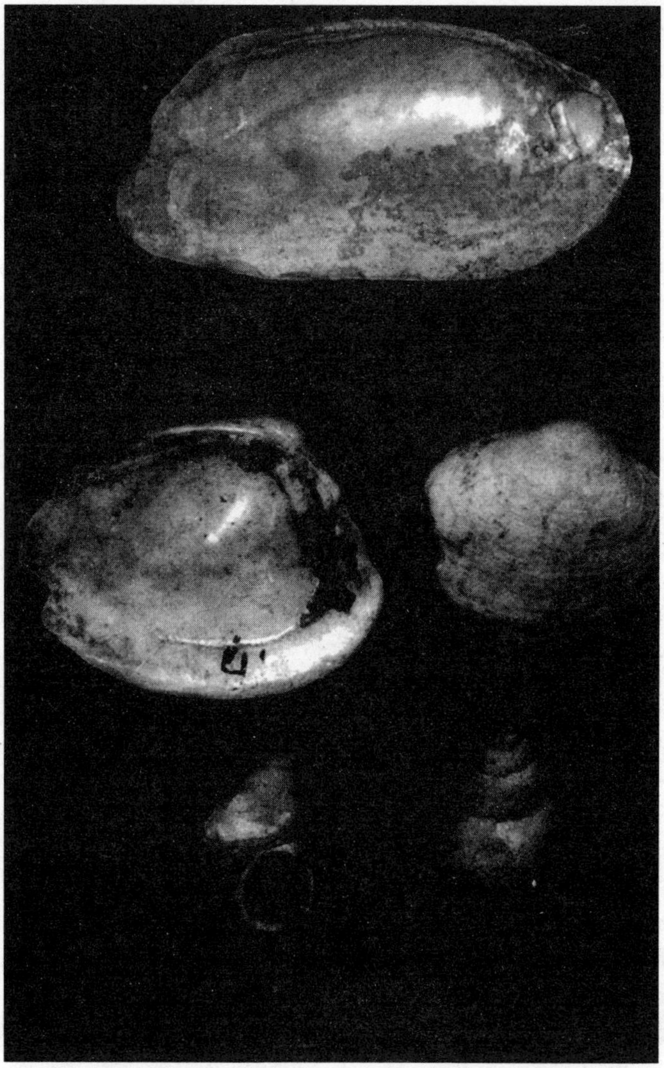

fig. 1: Fresh water shells from the Central Narmadā Valley. (top row: *Lamellidens* sp., middle row: *Parreysia* sp., bottom row: *Bellamya bengalensis*)

Modern representatives of these species inhabit sandy-muddy bottoms and banks of slow moving rivers, streams and lakes, and occur commonly in a majority of the Indian rivers. The bivalves are generally sessile as compared to the gastropods. Often, when rivers are in spate during monsoons, both bivalves and gastropods may be deposited on the river banks as a result of inundation.

Presence of a few complete shells of all the bivalve species with both valves intact shows that these were buried in their living condition. Further, the gastropods are also not very fragmented and abraided. Occurrence of such partially broken to complete shells indicates that deposition was *in situ* without much transportation, representing an autochthonous situation. The shells, in general, are not much rolled but are discoloured with a chalky white appearance. This reflects *in situ* modification of the shells by the overlying deposits which, in a way, have contributed to a better preservation of the shells.

The size of *Lamellidens, Parreysia* and *Bellamya,* when compared with similar modern specimens, reveals that the shells were in quite a mature stage of growth (about 1-2 yrs old). It further points out that favourable growth conditions like slow moving, less turbid water bodies and undisturbed and suitable muddy substratum were available to the shells. The tributaries of the river Narmadā, particularly the Sher and the Umer, seem to have been ideal for the growth of these species. However, slight local environmental fluctuations are reflected in the shells in the form of growth breaks visible on the shell surface of bivalves. The shell surface shows distinct growth breaks with the occurrence of growth lines. Growth line analysis in the case of marine shells has been useful in identifying the seasons in which growth bands have been deposited and also in estimating seasonality of the archaeological sites (Coutts, 1970; Koike, 1973). In the case of Narmadā shells, growth might have been interrupted by local environmental conditions like lowering of water table and drying up of local streams, turbidity, etc. In order to gain insight into this aspect, sectioning of shells and examination of growth lines is necessary which could not be attempted for the present material.

Attempts were also made to identify evidence of any human activity on the shells. This is absent although semi-fossilized vertebrate remains along with a few tools have been found in association with the shells. These associated cultural materials are mostly secondary in context while the shell assemblage is naturally occurring. More observations are necessary to explore this aspect of human use. So far, very few Upper Palaeolithic sites have revealed dietary use of freshwater shells, e.g. Dharampurī (Mishra, 1985).

Conclusion

The brief examination of fresh water molluscan shells from the Central Narmadā Valley has allowed some insight into the ecological and taphonomical history of the shells. It has helped to take into account the molluscan diversity in the region during the Pleistocene period. Occurrence of complete to partially broken shells of five species shows that they were deposited in their living state without getting transported for a long distance. These are naturally present and do not reflect any human activity. Fairly mature-sized shells, mainly the bivalves, show that suitable undisturbed growth conditions were available to the shells indicating that the Narmadā and its tributaries had plentiful water all round the year. This also suggests that moist climatic conditions prevailed which encouraged good vegetational growth. This can be further attested by the presence of different types of vertebrate fossils ranging from large mammals to small mammals and reptiles found existing in the area during Middle to Upper Pleistocene period.

Acknowledgement

I wish to thank Dr G.L. Badam for kindly permitting me to study the shell material collected by him from the Central Narmadā Valley.

References

Agrawal, D.P., & S. Kusumgar, 1974, *Prehistoric Chronology and Radiocarbon Dating in India,* New Delhi: Munshiram Manoharlal.

Badam, G.L., 1990, On a Partial Skull of *Crocodylus palaeindicus* from the Late Pleistocene Alluvium of the Central Narmadā Valley, Madhya Pradesh, India with Special Reference to Chronology and Palaeo-environment. Paper presented at the seminar on Environmental Change and Human Response (last two million years) at the Deccan College, Pune in December 1990.

Badam, G.L. & C. Grigson, 1990, A Cranium of Gaur, *Bibos gaurus* (Bovidae Mammalia) from the Pleistocene of India, *Modern Geology* 15: pp. 49-58.

Burleigh, R. & M.P. Kerney, 1982, Some Chronological Implications of a Fossil Molluscan Assemblage from a Neolithic Site at Brook, Kent, England, *Journal of Archaeological Science* 9: pp. 9-38.

Coutts, P.J.F. 1970, Bivalve-growth Patterning as a Method for Seasonal Dating in Archaeology, *Nature* 226: p. 874.

De Terra, H. & T. T. Paterson, 1939, *Studies on the Ice-Age in India and Associated Human Cultures,* Carn Institute: Washington Publication, p. 493.

Evans, J.G., 1969, Land and Freshwater Mollusca in Archaeology aspects, *World Archaeology* 1: pp. 170-83.

Koike, H., 1973, Daily Growth Lines of the Clam, *Meretrix lusoria*: A Basic Study for the Estimation of Prehistoric Seasonal Gathering, *Journal of the Anthropological Society of Nippon* 81: pp. 122-38.

———, 1977, *Estimation of Palaeotemperature by means of Growth Line Analysis,* The University Museum, University of Tokyo: Collected reprints, vol. 10, pp. 1-7.

Meighan, C., 1969, Molluscs as Food Remains in Archaeological Sites, in *Science in Archaeology* (Brothwell, D.R. & E.S. Higgs, eds.), London: Thames and Hudson, pp. 415-22.

Mishra, S., 1985, Early Man and Environments in Western Madhya Pradesh, Unpublished PhD Dissertation, Pune, University of Poona.

Patnaik, R., G.L. Badam & V. Sathe, 1995, Discovery of Micro-vertebrates from the Pleistocene Deposits of the Central Narmadā Valley, India, *Current Science* 68 (8): pp. 828-30.

Shackleton, N.J, 1970, Stable Isotope Study of the Palaeo-environment of the Neolithic site of Nea Nikomedia Greece, *Nature* 227: pp. 943-44.

Sparks, B.W., 1961, The ecological interpretation of Quaternary non-marine Mollusca, *Proceedings of the Linnean Society of London,* 172: pp. 71-80.

Subha Rao, N.V., 1989, *Fresh water molluscs of India,* Zoological Survey of India, Kolkata.

Theobald, W., 1860, On the Tertiary and Alluvial deposits of the Central Portion of the Nerbudda Valley, *Memoir Geological Survey of India* 2: pp. 279-98.

Tonapi, G.T., 1980, *Fresh Water Animals of India, an Ecological Approach.* New Delhi: Oxford and IBH Publishing Co.

Appendix III
Soil Chemical Analysis

Anupama Kshirsagar

CHEMICAL analysis of soil samples from two profiles, Devakachār (DVK) & Umeriā (UMR) from the Central Narmadā Valley was carried out. The samples were supplied by Dr G.L. Badam, in-charge of the IGRMS Narmadā Project. Earlier, several bone samples from various sections (before the commencement of the IGRMS Project) were also run for fluorine/phosphate analysis in order to get an idea about the chronological sequence of the beds (Kshirsagar and Badam, 1990). The results of the present analysis are given below in a tabular form (see Table 1).

Interpretation

pH of all the samples shows the neutral to slightly acidic nature. Most of the samples are gravels and organic carbon is very less, ranging from 0.13 per cent to 0.24 per cent with the exception of Black Cotton Soil (no. 1639) which contains 0.42 per cent organic carbon. Calcium carbonate in all these samples is also less, approximately ranging from 2 per cent to 6 per cent. There is not much variation or specific trend (increasing or decreasing) in silica, aluminium oxide or iron oxide in the profiles. With the exception of the sample from sandy unit (no. 1635) which contains 35.06 per cent SiO_2, 9.52 per cent Al_2O_3 and 10.08 per cent Fe_2O_3, all other samples contain 25 per cent to 31 per cent silica, 15 per cent to 20 per cent Al_2O_3 and 11 per cent to 14 per cent Fe_2O_3.

From this analysis it can be suggested that these sediments had their source in an area where the rainfall appears to have been significantly high as indicated by pH (neutral to slightly acidic), low silica content and fairly high Fe_2O_3 content. The second inference is that these sediments are derived from oxysols, i.e. strongly leached soils which might have been developed on the valley slopes of the Narmadā basin. There is no evidence of any development of soil in pedological sense in the upper profile at Devakachār. This supports the view that the sediments may have been deposited fairly rapidly in an alluvial plain environment without any significant time gap.

Reference

Kshirsagar, A. and G.L. Badam, 1990, "Biochronology and Fluorine Analysis of some Pleistocene Fossils from Central and Western India," *Bull. Deccan College Research Institute* 49: pp. 199-211.

Table 1

Results of chemical analysis of soil samples from IGRMS Narmadā Project

Lab No.		pH	% 0.C	% $CaCO_3$	% SiO_2	% Al_2O_3	% Fe_2O_3
1629	NMD Gugt DVK 3.45 m Mottled clay (Ist unit from below)	7.1	0.236	1.96	26.19	15.87	10.89
1630	NMD Gugt DVK 3.45 m Grey band (2nd unit from below)	7.0	0.126	1.96	24.98	14.60	11.66
1631	NMD Gugt DVK 3.45 m Loose pebbly gravel (3rd unit from below)	7.0	0.168	1.96	30.22	19.05	13.60
1632	NMD Gugt DVK 3.45 m sandy pebbly gravel (4th unit from below)	6.8	0.126	1.96	28.21	17.78	12.52
1633	NMD Gugt DVK 3.45 m sandy pebbly gravel with black bands (5th unit from below)	7.0	0.210	0.98	30.22	15.87	11.87
1634	NMD Gugt DVK 3.45 m from red clay horizon (6th unit from below)	7.1	0.210	5.88	27.00	20.95	13.84
1635	NMD Gugt DVK 3.45 m from sandy unit (7th unit from below)	7.0	0.189	--	35.06	9.52	10.08
1636	NMD Gugt DVK 3.45 m from consolidated sandy unit (8th unit from below)	7.1	0.168	4.9	28.21	15.87	13.84
1637	NMD Gugt DVK 3.45 m from shelly gravel (9th unit from below)	7.1	0.168	--	29.02	20.32	13.84
1638	NMD Lku UMR yellowish soil	7.0	0.168	1.96	31.03	19.05	12.42
1639	NMD Lku UMR Black Cotton Soil	6.9	0.42	3.92	24.98	19.05	12.96
1640	NMD Lku UMR Reddish sandy pebbly gravel	7.1	0.168	4.9	24.98	17.14	12.52
1641	NMD Lku UMR Blackish sandy pebbly gravel	7.1	0.147	3.92	28.21	15.87	13.60
1642	NMD Lku UMR yellowish sandy pebbly gravel	6.8	0.210	4.9	24.18	15.87	13.07
1643	Behind Gugt (soil surrounding the fossil)	7.1	0.126	3.92	24.98	17.78	11.87

Appendix IV

Animal Depictions in Rock Art and Palaeoecology

A Case Study at Bhīmbeṭkā, Madhya Pradesh, India*

G. L. Badam and *V. G. Sathe*

Abstract

RESEARCH on Rock Art in India has gained momentum during the last two decades or so. However, the studies have mainly been restricted to identification of animals (sometimes made wrongly), styles of animal representations and dating methods. The present paper is the first attempt to study the ecological aspect of animals depicted in the rock art in India. In general the animals depicted are elephant, bison, bull, buffalo, antelope, various deer, rhinoceros, tiger, mongoose, rat, porcupine, wild boar, tortoise, etc. In addition, some composite animals (e.g. boar with horns as at Bhīmbeṭkā) are depicted in various galleries.

A case study undertaken of the Bhīmbeṭkā area in Madhya Pradesh with the help of living analogies has helped in reconstructing more accurately the palaeoecology of the area during Mesolithic to historic times. Mesolithic is a phase of faunal turnover in India which gave rise to progressive forms, most of them continuing even until the present times.

Practically all the representative members of the forest types such as tiger, elephant, medium to large-sized ungulates, etc. are depicted. Barring some predators the fauna represents grazing and/or browsing and mixed feeding habits. The consistent flow of floral biomass contributed by woody plants, shrubs and tall grasses was a prerequisite to maintain such an ideal ecosystem.

In the present paper, the ecology of each animal group is discussed chronologically and a model for different animal communities in various tiers of the ecological pyramid for Bhīmbeṭkā suggested.

---

* Revised version of *Rock Art — The Way Ahead*, General Editor : Shirley — Ann Pager (Occasional SARARA publication no. 1), pp. 196-208, 1991.

Introduction

Paintings on bare rocks by ancient man provide both a glimpse of their world and insights into the evolution of human conceptual abilities. Beginning in India and Spain in the last century (Cockburn, 1899), a large number of rock painting sites have been discovered which provide excellent records of early human culture.

One such site is Bhīmbeṭkā (22° 54' 58" N: 77° 36' 59" E) in the District of Raisen of Madhya Pradesh state which dates from early Mesolithic to historic times. Animal depictions are the predominant subject matter throughout this time period (*Pl.* 1), while human figures appear slightly later in the Mesolithic.

Rock art is relevant to biological and palaeobiological studies in that the paintings depict a varied fauna and flora which can be compared to the large number of animals and plants currently in the vicinity of Bhīmbeṭkā and other rock shelters. The present paper is the first attempt to reconstruct the palaeoecology of a site based upon faunal depictions in rock paintings. The need for proper identification of animals depicted in paintings for a more accurate palaeoecological reconstruction is stressed.

The reference point of Bhīmbeṭkā lies about 2 km south of a small tribal village of Bhiyanpur in Raisen District. The site is 40 km south of Bhopal and 30 km north of Hoshangabad, and almost equidistant between Ubaidullaganj and Barkheṛā (*fig.* 1). Bhīmbeṭkā is the name of a hill in the Western Vindhyas with a height of 600 m above mean sea level (MSL) and extending about 2 km in both length and breadth. The hill is oriented north-east-south-west and is characterized by steep slopes on the eastern and southern sides. Bhīmbeṭkā and its vicinity are drained by the Bān Gaṅgā, the Gupta Gaṅgā and the Pandapur, three perennial springs which are located towards the north-eastern foot of the hills. In addition, a number of small rivulets and creeks on the southern side join smaller rivers to flow into the Narmadā river.

The history of the Bhīmbeṭkā area was studied by W. Kincaid (1888), who referred to the hills of Bhīmbeṭkā as a Buddhist site. However, the significance of the area in the field of rock art was brought to light in 1958 by the late Dr V.S. Wakankar. After a lapse of about one and a half decades, excavations of the floor of shelters (III-F-24 and II-B-33) were undertaken by him which revealed evidence of Acheulian and Mesolithic occupation in the area. Systematic excavations were resumed on a large scale by Dr V.N. Misra at III-F-23 (*Pl.* 2) (Misra, 1985), bringing to light a sequence of artifacts of late Acheulian to late Mesolithic origin. Some more excavations, conducted at III-F-13 and II-B-33, helped reveal the evolution of Mesolithic typology. Acheulian, Middle Palaeolithic and Mesolithic artefacts excavated by Wakankar in eight shelters have highlighted the evolutionary aspect of tool technology in isolation in a more or less homogeneous environment. The association of this tool technology with the rock art evoked tremendous attention regarding the

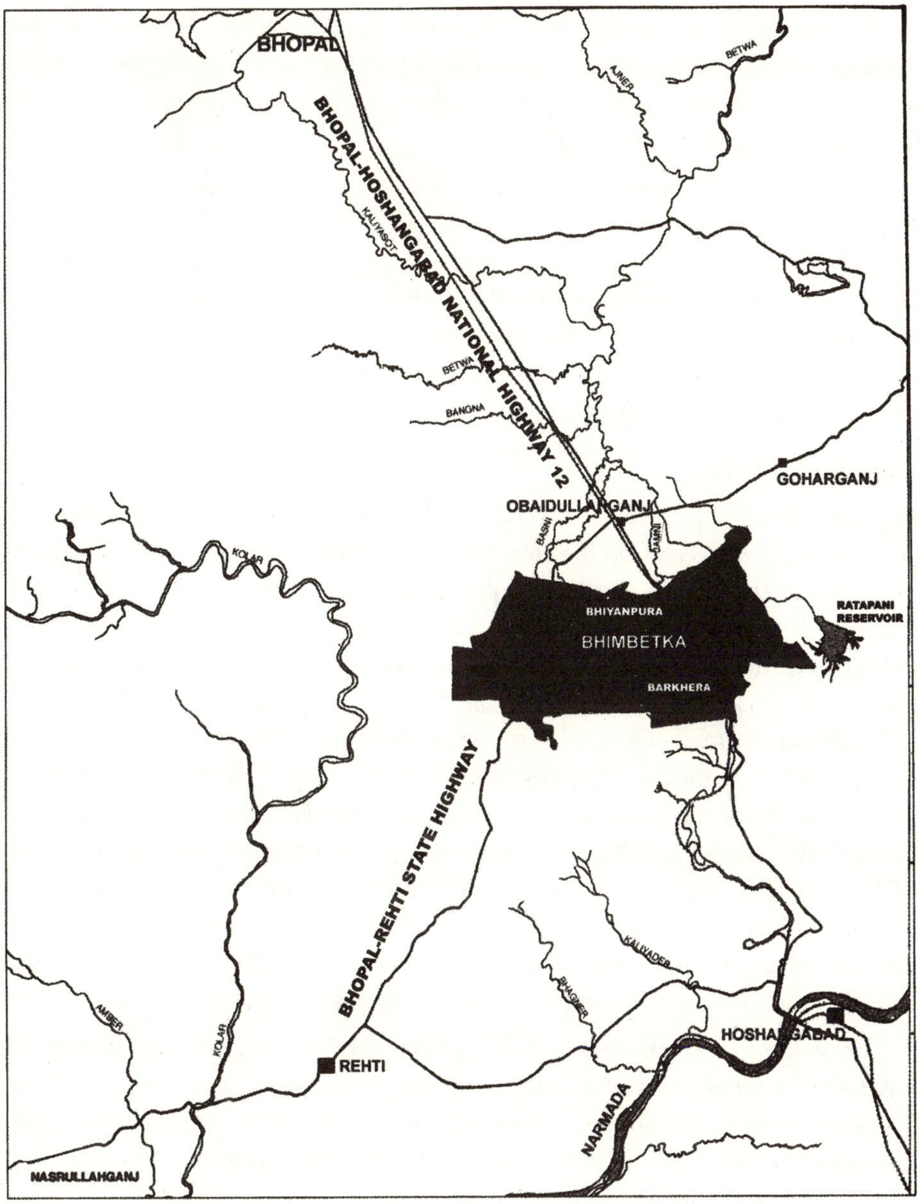

fig. 1: Location of Bhīmbeṭkā, District Raisen, M.P.

antiquity of the paintings (Wakankar 1973, 1975; Wakankar and Brooks, 1976; Neumayer, 1983; Mathpal, 1984; Chakravarty, 1984).

Geology and Climate

Geologically, these hills are composed of sandstone of Upper Vindhyan System, dated at about 600 million years BP. The rocks, being sedimentary in origin, exhibit distinct layering and ripple marks. At places, sandstone has metamorphosed into orthoquartzite. Weathering and natural factors such as sun, rain, wind, acid etching and temperature fluctuations as well as penetration by tree roots and insects (bioturbation), have decayed softer sediments to produce rocks with different and uncommon shapes.

In certain cases, huge horizontal rocks have detached themselves from the rocks beneath and have fallen to rest at various angles. Thus, the formation of most of the shelters at Bhīmbeṭkā is attributed to the enlargement of cavities through physical and chemical weathering.

The hills of Bhīmbeṭkā and the surrounding region form a watershed zone. A tropical monsoon-type climate predominates in the area, with an annual rainfall of 1.026 mm. More that 90 per cent of the rainfall is received between July and September. October to February is the winter period, with January being the coldest month. Summer is quite harsh and spans between March and May. Bhīmbeṭkā is actually situated on the Malwa plateau and, hence, the nights are characteristically cool during the summer.

Flora and Fauna

Dry deciduous vegetation is the chief forest type that covers the hills in the area. A large variety of edible flowers, fruits, seeds and tubers available here provide subsistence to the local tribal population throughout the year. The individual species of trees include Achar (*Buchanania latifolia*), Amaltās (*Cassia fistula*), Āonlā (*Phyllanthus emblica*), Bahera (*Terminalia belerica*), Bamboo (*Dondrocalamus strictus*), Bel (*Aegle marmelos*), Ber (*Zizyphus jujuba*), Dudhai (*Wrightea tomentosa*), Gondi Lasora (*Cordia mxya*), Gurjan (*Garuga pinnata*), Haldu (*Adina cordifolia*), Hārsingār (*Nyctanthes arbortristis*), Jāmun (*Eugenia jambolana*), Kanchan (*Bauhinia variegata*), Kurchi (*Holharrena antidysenterica*), Khair (*Acacia catachu*), Khajur (*Phoenix sylvestric*), Khirni (*Mimusops hexandra*), Mahuā (*Madhuca latifolia*), Palāś (*Butea frondosa*), Sagaun (*Tectona grandis*), Asan (*Terminalia tomentosa*), Semel (*Bombax malabaricum*), Shisham (*Delbergia sissoo*), Tendu (*Diospyros tomentosa*), and Tūn (*Cardela toona*). Wild varieties of tubers are plentiful during the rainy season and the beginning of winter.

Over the years, the diverse forest at Bhīmbeṭkā has provided an excellent habitat for a large number of carnivores, herbivores and various smaller fauna. The animal taxa include tiger (*Panthera tigris*), Panther (*Panthera pardus*), sloth bear (*Melurus ursinus*), hyaena (*Hyaena hyaena*), wolf (*Canis lupus*), fox (*Vulpes bengalensis*), jackal (*Canis aureus*), langur (*Presbytis entellus*), boar (*Sus scrofa*), nīlgāi (*Boselephus tragocamelus*), sambar (*Cervus unicolor*), swamp deer (*Cervus duvauceli*), chītal (*Axis axis*), Chinkārā (*Gazella gazella*), barking deer (*Muntiacus muntjak*), porcupine (*Hystrix indica*), hare (*Lepus nigricollis*) and pangolin (*Manis crassicaudata*). However, elephant, rhinoceros, lion, gaur and water buffalo, all predominant in the paintings, are not found in the area today.

Age and Chronology of the Paintings*

The antiquity of rock paintings has always been a matter of dispute since their initial discovery in India and Spain during the second half of the previous

---

* Scientific dating techniques link AMS, OSL etc. have recently been initiated in Bhīmbeṭkā samples which hope to put the chronology on a firm footing.

century. However, dating has achieved a certain amount of legitimacy by systematic classification, study of the subject matter, technique, and above all, the confirmed association with radiometrically dated pre-historic cultural remains. The paintings at Bhīmbeṭkā have been documented and studied carefully in an archaeological context. Thus, the rock art has been assigned to the timespan from the Mesolithic to the Chalcolithic and historic times, supported by the following C14 dates :

Mesolithic : 8000 BCE

*Transitional : 3000 BCE

Historic : 600 BCE to CE 500-700

(Mathpal, 1984 & Wakankar, 1984)

Taking into consideration the subject matter, technique, colour and the superimposition of various styles on one another, the paintings have been subdivided into several periods and phases.

The first period is associated with Mesolithic, the second period (considered transitional) with the Chalcolithic, and the third period with an historical origin (Mathpal, 1984), as follows:

Period 1 : (A)
(B)
(C) — Mesolithic
(D)
(E)

Period 2 : (F) Transitional

Period 3 : (G)
(H) — Historic
(I)

For a better understanding of contemporary fauna, all the zoomorphic representations are taken into account with respect to the phase-wise distribution.

Period 1

PHASE A

This is the earliest phase (Early Mesolithic) of the paintings, being characterised by 26 large animal figures outlined and partially filled in colour in geometric designs. The absence of human figures is confined to this phase only. Animals depicted include buffalo, leopard, cow and elephant.

---

* Transitional on the basis of style, subject matter and features related to paintings and to tool assemblages and chronologically associated with Chalcolithic.

PHASE B

Human as well as animal figures total 1,167 in this phase. Buffalo, cow, ox (*Bos indicus*), gaur, nīlgāi, chiṅkārā, black buck, goat, sambar, swamp deer, barking deer, boar, rhinoceros, dog, wolf, hyaena, bear, rabbit, monkey and langur are generally depicted in this phase. However, chītal is the most profusely painted animal, possibly indicating an increase in its population. Animals are generally shown in a state of fright or as running and jumping.

The artists may have been attempting to show the non-sedentary nature of these species and the ungulates' response to the presence of predators.

PHASE C

The animal and human figures become slightly larger with a marked increase in the number of anthropomorphic representations. The depictions total 591 figures, the majority (83) being of various types of deer followed by cow, ox, goat, antelope, buffalo, nīlgāi, leopard, elephant, boar, tiger and lion. Hunting scenes are common and hand prints of man appear for the first time.

PHASE D

The number of pictures reduces to 280 (183 human) in this phase, which is characterized by the beginning of the stylized portrayal of living objects. Animals portrayed include ox, chiṅkārā, deer, dog, elephant and buffalo.

PHASE E

There are a total of 70 paintings of buffalo, chītal, sambar, swamp deer, goat, boar, chiṅkārā and cow in this phase. The human figures total 11.

Period 2

PHASE F

This constitutes a total of 172 paintings. Animals which predominate are ox, dog, chītal, boar, tiger, cow, nīlgāi, elephant and leopard. The transitional aspect of this phase is attributed to the domestic nature of the animals which appear quite familiar with human figures. The style of the paintings indicates the beginning of domestication. The decoration of animals, particularly of the horn tips, seems unique in this phase. Sometimes, dogs and oxen are shown next to tiny, child-like figures. This is of particular interest because these were the first animals to become domesticated during the early period of civilisation. This phase has been chronologically associated with the Chalcolithic period (Wakankar & Brooks, 1976; Mathpal, 1984).

Period 3

PHASES G, H & I

(Phases G, H and I have been treated together as the data in individual phases is meagre).

Significantly, the recurring appearance of horse images places these phases in the Historic period. There is a marked increase of human figures and "domesticated looking" animals, suggesting a symbiotic relationship.

Conflict and competition (an integral aspect of a developing society) are also suggested by depictions of horsemen armed with spears, bows and arrows, swords, shields and other metal weapons. These paintings lack accurate anatomical portrayal and naturalism.

Phases G, H and I have a total number of 1424, 1374 and 1056 paintings respectively. Portrayal in phase G of elephant, laṅgūr, monkey, dog, sambar, ox, cow, chītal, camel, goat, giraffe (?), nīlgāi, swamp deer, boar, chiṅkārā, rabbit, fish, lizard and birds* is clear. These taxa continue to be the subject matter in phase H, with the exception of leopard, tiger, chītal, buffalo and goat. Paintings of phase I appear close to the folk art. Human, animal and fragmented, unidentifiable figures total 176, 207 and 16 respectively. The remaining 657 figures have been classified as decorative designs and/or some sort of writings (Mathpal, 1984).

Ecological Hypothesis

A well documented pictorial record of different animals is now available from Bhīmbeṭkā as direct evidence of their existence in and around the site. However, the style, technique and superimposition tend to suggest that all the animals depicted did not live simultaneously at this site. The absence of certain animals on the shelters does not necessarily indicate their total physical absence from the area while the painters were at work. At the same time, depictions of some animals might have been an outcome of their encounter elsewhere, outside the vicinity of Bhīmbeṭkā. Some of the depictions may even be composite figures, the result of the artist's imagination.

The ecological conditions prevailing while the painters were at work, can be largely revealed by studying the ecology of those animals (predominantly mammals) represented in the paintings that are present today. In this context, exhaustive treatment is given on the habitational preferences of individual animals and the factors which control their movements. Archaeologically, the time period involved of the fauna depicted here, is short and does not go beyond the beginning of the Holocene.

Palaeontologically, this period marks the faunal turnover in India, giving rise to some progressive forms. However, a comparative study of the depicted animals with the help of modern analogues is more meaningful even though many of the archaic fauna no longer exist today. It may be noted that even today Bhīmbeṭkā enjoys a great variety of floral and faunal wealth, which is well documented.

Ecologically, the faunal biota represented in the paintings can be divided into the following palaeo-communities:

* A few scholars opine that some of the bird paintings are those of ostrich.

1. Stream and streambank community
2. Megaterrestrial community
3. Arboreal community

STREAM AND STREAMBANK COMMUNITY

Turtle, crab and some varieties of fish are a few representatives of this community. The physical setting of the area characterizes three perennial aquatic sources (the Bāngaṅgā, the Gupt Gaṅgā and the Pandapur) which could be inhabited by these small creatures. In all probability, these animals were a potential source of food for early man inhabiting these shelters. At times, their portrayal may even be symbolic, representing the presence of water or a related environment.

MEGATERRESTRIAL COMMUNITY

This is the most predominant community represented throughout the paintings of all periods. The members are diverse and include large to small-sized herbivores with mixed feeding habits of grazing and browsing (elephant, buffalo, chītal), carnivores (tiger, leopard), scavengers (hyaena) and small mammals (hare, mongoose, squirrel and porcupine). Depictions of such a diverse biota undoubtedly reveals floral heterogeneity, the composition of which was primarily dependent upon the climate and soil of the area. Chemical analysis of the soil from one of the shelters (BTK III-F-23) has indicated that throughout the Stone Ages the region experienced a humid climate with a mean annual rainfall of more that 100 mm (Rajaguru, 1984). The consistent depiction of a rich fauna in the paintings suggests that such climatic conditions with minor fluctuations might have persisted for a long time period.

The large bovids like bison, nīlgāi and to some extent water buffalo are well known for their adaptability to macro-environments. Because grazing is the sole mode of foraging, these animals can be recognized as grassland traverses (a community group), making best use of perennially available grasslands. Those areas covered with primary and secondary forests might have controlled their movements. Water bodies which are usually found in open grasslands in humid climates, certainly fulfilled the basic water requirements of these animals.

However, the water buffalo (*Bubalus bubalis*) occupies a somewhat unique habitational environment which is essentially associated with permanent water bodies and a staple source of grass to fulfil all its ecological requirements. The semi-aquatic nature of buffalo is further indicated by the fact that most of its time is spent wallowing in water, which acts as an important factor for normalization of its environmental temperature (Fahimuddin, 1975). Buffalo, with its wide horns and massive structure, is one of the first animals to be painted at Bhīmbeṭkā.

Rhinoceros (*Rhinoceros unicornis*) is depicted as being hunted or in motion. These paintings are dated to the Mesolithic on the basis of style and use of colours. Rhinoceros survives in an environment which is widely interspersed with grassland and swampy areas. These animals frequent forests. However, being chiefly a grazer, foraging occurs mostly in grassy plains (Prater, 1971). The movements of rhinoceros are generally confined to the fringes of bushy, moist grasslands which border primary and secondary forests.

Cervids generally thrive in a different ecological setting, one which is represented by swampy grasslands interspersed with water bodies, particularly in the dry season. Cervids like sambar (*Cervus unicolor*), bārāsiṅghā (*C. duvauceli*), chītal (*Axis axis*) and barking deer (*Muntiacus muntjak*) must have thrived at Bhīmbeṭkā since all their ecological requirements were easily met. However, bārāsiṅghā represents a somewhat unique ecological range and hence, needs special mention. Three sub-species exist today in different ecological zones in India. These are *C.d. duvauceli*, *C.d. ranjitsinhi* and *C.d. branderi*, of which the latter inhabits Central India. Ecological studies carried out on bārāsiṅghā at Kanha National Park, Dist. Mandla of Madhya Pradesh indicate semi-evergreen and wet evergreen forests as the habitational zones of this animal (Martin, 1977; Groves, 1982). There are three main factors known to control their movements: scarcity of water, abundance of grasscover and climatic conditions. Browsing is occasional but the animal is practically dependent upon a grass diet throughout the year. In general, the climatic changes induce seasonal movements, the drier conditions bringing them close to aquatic-swampy-grassy sheets. Fringes of the forest and better drained areas attract bārāsiṅghā with the advent of monsoons, while the drier conditions may direct them from glades to woods (Martin, 1977). In contrast to bārāsiṅghā, the chītal and sāmbar are found less frequently in swamps and frequently traverse woods. However, they do compete with bārāsiṅghā by foraging in potential food yielding areas. Smaller members of bovids like *Antilope cervicapra* and chiṅkārā can often be spotted grazing in association with these cervids and are not distinct from other grazers in feeding habits.

Wild boars are one of the most frequently painted animals at Bhīmbeṭkā. Horns, nasal horns, massive size, x-ray like paintings, and unique body decorations have rendered most of them composite animals. Hunting of wild boars by stick-like human figures is common in the Mesolithic and early Chalcolithic paintings. Dense forests with sufficient tubers, shrubs, and aquatic sources could have been frequented by boars in a humid, wet Mesolithic phase. Recurring reference to boar sheds light on the subsistence strategy of the painters. In addition to vegetable foods, it is possible that pig was one of the most staple and consistent sources of food. A deified status for the animal (Neumayer, 1983), however, remains enigmatic and needs to be substantiated by further studies on religion and symbolism of early populations.

Elephant is the largest herbivore among the depicted animals at Bhīmbeṭkā. The physiological and anatomical characteristics of the animal indicate that deriving maximum energy from the plant resources is a prerequisite for its very survival. In this regard, elephant stands out as the largest consumer of vegetation. Whether elephant is primarily a grazer or a browser, has been a matter of debate. On the basis of the high proportion of grass in its diet (Buss, 1961; Mckay, 1973), grazing appears to be the main feeding habit of the Asian elephant (Olivier, 1978). However, based on its nutritional requirements, Sikes (1971) and Laws et al. (1975) have emphasized the need for elephant to consume a higher proportion of plants obtained by browsing. The habitat of elephant comprises grassland and forest cover in roughly equal proportions. It has been observed that it spends more time at the forest and grassland interface (Mckay, 1973).

This suggests two types of opportunistic feeding tendencies: in pastures in close proximity to forest and foraging in ecotone areas, which supply a variety of potential food plants during all given feeding periods. This mode of foraging has been nutritionally more significant as grass in wet seasons and browse in dry seasons are two broadly recognized diets which are likely to substantiate their demands in all seasons. In other words, the optimal diet of the elephant appears to be a reasonable mix of browse and grass, depending upon the seasons and the types of vegetation available to it (Sukumar, 1989). This tends to indicate a similar ecological setup at Bhīmbeṭkā that formed a fundamental niche, promoting long-term survival of elephant in the area. Olivier states that ". . . the mere presence of an elephant in the ecosystem is evidence of considerable floral diversity" (1982: 297).

Bear is an omnivorous animal with a behavioural pattern frequently similar to that of humans. It is however, predominantly frugivorous. Hilly terrain with dense, woody, rich, fruit-yielding forests and rock shelters, probably provided an excellent niche at Bhīmbeṭkā. It is possible that the painters often encountered bear.

Life of a predator in a given area depends solely upon the availability of large herbivore biomass. This is particularly true in the case of tiger, lion, panther and leopard who are strong and successful predators. And to support a large quantity of prey animals (e.g. herbivores/small carnivores/monkeys), a perennial source of vegetation is the fundamental requirement. A proper maintenance of the food chain necessitates floral diversity, where the grass plants constitute the base of the biomass pyramid. A distinct decrease in the consumer levels at each stage forms the "tiers" of the pyramid (*fig.* 2). Characteristically (such as at Bhīmbeṭkā), one or a few predators are placed at the top. Hence, the presence of predators like tiger, lion and panther (as evidenced by the paintings), indicates the existence of a well balanced ecosystem, where every member of the food chain had a deserving share.

Scavengers like jackal, wolf, fox and hyaena were probably the second largest consumers of meat in the area. These animals subsist on the leftovers of prey killed by larger predators. Their depiction fills up the remaining aspect of food chain.

Pl. 1: Drawings of animals on a panel in a Bhīmbeṭkā rock shelter.

Pl. 2: A view of Bhīmbeṭkā rock shelter, III — F-23.

Smaller animals painted throughout the Mesolithic to the Historic period include mongoose, rat, porcupine and lizard which are generally associated

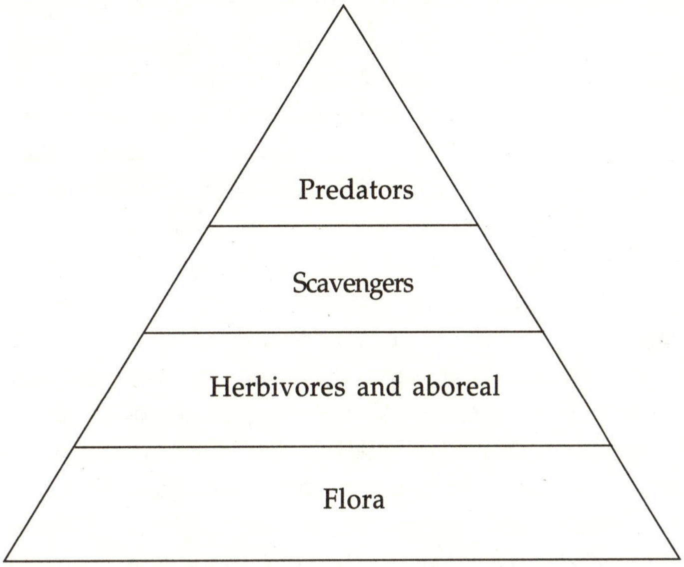

fig. 2

Proposed Palaeoecological Model at Bhīmbeṭkā

with such an environment. The behavioural niche of the animals is so wide that their mere presence does not point to any particular ecological set-up.

> Flora is a fundamental requirement which forms the base of the pyramid of biomass. A successive reduction in consumer level at each stage forms the "tiers" of the pyramid. Predators are located on the top, denoting their supremacy over the food chain in terms of an ideal ecosystem. Scavengers fill up the remaining aspect of food chain. They are the second largest consumers of meat after the powerful predators. They also live on the leftovers of animals killed by predators.

ARBOREAL COMMUNITY

Macaque (*Macca mulatta*) and laṅgūr (*Presbytis entellus*) represent the arboreal community depicted in the paintings at Bhīmbeṭkā. The foraging preferences and the ecological niche of these primates is known to be in secondary forest types, where there is a constant and plentiful supply of foodstuffs. The abundance of plant species in tropical rain forests provides considerable scope for the proliferation of the monkeys thriving in such areas. Thus the field niche of primates may result from choice and adaptation. The forest around Bhīmbeṭkā might have been a choice niche for them.

Discussion

The human response to the biological environment during pre-historic times has been well demonstrated by the elaborate paintings at Bhīmbeṭkā. Animal representation is systematic and explicit. In contrast, human figures are stick-like with minimal identification to denote them anthropomorphic. Very few figures (animal or human) are stationary and animals are shown as an integral part of human activity, suggesting a close relationship between the two. There also existed a possible peaceful co-existence within the human population. The portrayal of animals, whether being hunted, shown in herds, or drawn in composite, convincingly reveals that the faculty of expression had attained great heights in the primitive population at Bhīmbeṭkā in Mesolithic times. Floral heterogeneity, in terms of primary to secondary forests with intervals of open savannah grasslands, is inferred and the phenomenon was consistent from prehistoric to historic times. All the components of an ideal ecosystem (predators, scavengers, browsers, grazers and monkeys, as well as smaller animals) indicate the presence of a consistent flow of floral biomass comprising woody plants, shrubs, and tall grasses in the vicinity of Bhīmbeṭkā. Early man also consumed plants, fruits and tubers, foods which are important to the tribal population in the area today.

A palaeoecological interpretation at Bhīmbeṭkā can be closely linked with the palaeoecology of the Narmadā valley. A geological and palaeontological study of the Central Narmadā Valley over the last two decades (Badam, 1979, 1984) indicates that the Narmadā and its tributaries frequently changed course due to an overall aggradational tendency of the drainage system during late Pleistocene times. The Central Narmadā, because of its geographical location, did receive comparatively large amounts of rainfall. This climatic condition led to frequent floods in the agrading Narmadā and its tributaries. Faunal evidence suggests that the vegetation during this period was of the savannah type with several backswamps and water pools.

The Narmadā was rejuvenated in the early Holocene. The ground water was reduced to modern levels which resulted in a number of swamps and backswamps in this environment. At the same time, the early Holocene was characterized by increased rainfall and vegetation density, resulting in the receding of open spaces. This provided an ideal habitat for herbivorous ungulates, proboscideans and reptiles in the area. The carnivores and other animals which are represented in the Pleistocene by fossils and in the Holocene by rock paintings, were probably concentrated in the galleria forests near the banks of the rivers.

A similar situation must have prevailed in the Bhīmbeṭkā area, where Early Man witnessed all this and the rich diversity of animals in this ideal habitat. Even though it is difficult to visualize accurately the man-animal adaptations in the area, early man did express his artistic ability in the form of rock paintings.

In the end, it may be stressed that a proper identification of the animals in the paintings is a matter of great concern for a palaeontologist. This demands judicious interpretation of the portrayals, whether anthropomorphic or zoomorphic. It must be borne in mind that proper identification forms the very data base for theory building and any palaeoecological model building without proper identification of animals can lead to erroneous interpretations.

Acknowledgement

We are grateful to Professor S.N. Rajaguru (Deccan College, Pune) for a perusal of the manuscript and for his suggestions. Constructive criticism on an earlier draft of the paper by Dr Donald C. Dunbar (Visiting Scientist from U.S.A.) is gratefully acknowledged. This paper was intended to be presented in an International Seminar "Rock Art — The Way Ahead" sponsored by South African Rock Art Research Association (SARARA) during August 1991 in Natal, South Africa. GLB is grateful to Shirley-Ann Pager, the organizer of the seminar for the interest shown in this work and the encouragement given from time to time which helped in completion of the manuscript for inclusion in the publication schedule.

References

Badam, G.L., 1979, *Pleistocene Fauna of India*, Deccan College, Pune.

———, 1984, "Pleistocene Faunal Succession of India," in: Whyte, R.O. (ed.), *The Evolution of East Asian Environment*, University of Hong Kong, pp. 746-75.

Buss, I.O., 1961, "Some Observations of Food Habits and Behaviour of the African Elephant," *Journal of Wildlife Management* 25: pp. 131-48.

Chakravarty, K.K. (ed.), 1984, *Rock Art of India: Paintings and Engraving*, Arnold- Heinemann Publ., N. Delhi.

Cockburn, J., 1899, "Cave Drawing in the Kaimur Range, North-West province," *Journal of the Royal Asiatic Society* 51: pp. 89-97.

Fahimuddin, 1975, *Domestic Water Buffalo*, Oxford and IBH Publishing Co., New Delhi.

Groves, C.P., 1982, "Geographical variation in the Barasingha or Swamp Deer," *Journal of Bombay Natural History Society of India* 79(3): pp. 620-29.

Kincaid, W., 1888, "Rambles among ruins in Central India," *The Indian Antiquary*, 17: pp. 348-52.

Laws, R.M. Parker, I.S.C. and Johnstone, R.C.B. 1975, *Elephants and their Habitats*, Oxford: Clarendon Press.

Martin, C., 1977, "Status and Ecology of the Barasingha (*Cervus duvauceli branderi*) in Kanha National Park," *Journal of Bombay Natural History Society* 74(1): pp. 61-132.

Mathpal, Y., 1984, *Prehistoric Rock Paintings of Bhīmbeṭkā, Central India*, Abinav Publication, New Delhi.

Mckay, G.M., 1973, *Behaviour and Ecology of the Asiatic Elephant in South Eastern Ceylon, Smithsonian Contributions to Zoology*, 125.

Misra, V.N., 1985, "The Acheulian Industry of Rock Shelter III-F-23 at Bhīmbeṭkā, Central India," In: Belwood, Peter & V.N. Misra, eds., *Recent Advances in Indo-Pacific Prehistory*, Oxford and IBH Publishing Co. : New Delhi, pp. 35-47.

Neumayer, E., 1983, *Prehistoric Indian Rock Paintings*, New Delhi, Oxford University Press.

Olivier, R.C.D., 1978, *On the Ecology of the Asian Elephant* (*Elephant maximus* Linn), *with particular reference to Malaya and Sri Lanka*, PhD Thesis, University of Cambridge.

———, 1982, "Ecology and Behaviour of living elephants: Bases for assumptions concerning the extinct woolly mammoths," in: Hopkins, D.M.; Matthews, J.V. Jr.; Schweger, C.E. & Young, S.B., eds., *Palaeoecology of Beringia*, Academic Press, pp. 291-305.

Prater, S.H., 1971, *The Book of Indian Animals*, Bombay Natural History Society and Prince of Wales Museums, Western India, Bombay.

Rajaguru, S.N., 1984, "Mineralogy and Palaeo-environment of Bhīmbeṭkā Rock Shelter, Madhya Pradesh" in: Chakravarty, K.K., ed.,) *Rock Art of India*, Arnold-Heinemann Publishers, New Delhi, pp. 171-76.

Sikes, S.K., 1971, *The Natural History of the African elephant*, New York, Elsevier.

Sukumar, R., 1989, *The Asian Elephant: Ecology and Management*, Cambridge, Cambridge University Press.

Wakankar, V.S., 1973, *Painted Rock Shelters of India*, Ph.D. Thesis, University of Poona.

———, 1975, "Bhīmbeṭkā — The Prehistoric Paradise," *Prachya Pratibha* 3(2): pp. 7-29.

———, 1984, "Bhīmbeṭkā and Dating of Indian Rock Paintings," in: Chakravarty, K.K., ed., *Rock Art of India*, Arnold-Heinemann Publishers, New Delhi, pp. 44-56.

Wakankar, V.S. & Brooks, R.R.R., 1976, *Stone Age Painting in India*, D.P., Taraporevala & Sons Co. Bombay.

Bibliography

Agarwal, D.P. and S. Kusumgar, 1975, "Radiocarbon dates of some Late Quaternary samples," *Curr. Sci.,* 44(5), pp. 39-53 and 149-51.

Agrawal, D.P., B.S. Kotlia and S. Kusumgar, 1988, Chronology and significance of the Narmadā formations, *Proc. Ind. Nat. Sci. Acad.* 54 (A), pp. 418-22.

Ahsan, S.M.K., 1993, *A Study of Palaeolithic site Formation Process in Subhumid Environment of Central India with Special Reference to Samnapur Palaeolithic Site M.P.*, PhD Dissertation, Pune: University of Poona.

Azzaroli, A. and G. Napoleone, 1982, "A Magneto-stratigraphic Investigation of the Upper Siwaliks near Pinjor, India," *Rivista Italiana di Palaeontologia* 87 (4) : 79, pp. 39-62.

Badam, 1979a, "Quaternary Palaeontology of the Central Narmadā Valley and its Implications in the Prehistoric Studies," *Geol. Surv. Ind. Misc.* Pub. 45: pp. 311-20.

———, 1979b, *Pleistocene Fauna of India*, Pune: Deccan College.

———, 1982, "Biostratigraphy of the Central Narmadā Valley — a Reappraisal." in : R.K. Sharma (ed.), *Indian Archaeology New Perspectives,* Agam Kala Prakashan, Delhi, pp. 38-47.

———, 1984, "Pleistocene Faunal Succession in India," in : Whyte, R.O. (ed.), *The Evolution of East Asian Environment,* Centre of Asian Studies, Univ. of Hong-Kong, 2, pp. 746-75.

———, 1985, *"The Late Quaternary Fauna of Inamgaon,"* in V.N. Misra and Peter Bellwood (eds.), *Recent Advances in Indo-Pacific Prehistory,* Oxford and IBH Co., New Delhi, pp. 413-15.

———, 1988, *Quaternary Faunal Succession of India, Three decades of developments in Palaeontology and Stratigraphy in India,* 2, G.S.I. Special publication 11, pp. 277-304, Hyderabad.

———, 1989, "Observations on the Fossil Hominid Site at Hathnorā, Madhya Pradesh," in: *Perspectives in Human Evolution* (Ashok Sahni and Rajan Gaur, eds.), pp. 153-71, Delhi: Renaissance Publishing House.

———, 1990, Taphonomic Implications as Reflected in Large Vertebrates from Quaternary sites in Western India, 6th International Conference of the International Council for Archaeozoology, Simithsonian Institution, Washington D.C. (Abstract).

———, 1996, "Taphonomy as a tool in interpreting Fossils and their modes of occurrence," in Margbandhu, C. and K.S. Ramachandran (eds.), *Spectrum of Indian Culture, Prof. S.B. Deo Felicitation volume,* Agam Kala Prakashan, Delhi.

Badam, G.L. and C. Grigson, 1990, "A Cranium of Gaur, *Bibos gaurus* (Bovidae, Mammalia) from the Pleistocene of India," *Modern Geology,* 15, pp. 49-58.

Badam, G.L. and S.L. Jain, 1988, "First Record of *Bos namadicus* from Middle-Late Pleistocene deposite of Lower Godavari, near Servaipet in Karimnagar District, Andhra Pradesh," *Jour. Pal. Soc. India,* 33, pp. 97-104.

Badam, G.L. and M.D. Kajale, 1977a, "First record of a fossil *Hexaprotodon* from the Ghod Valley, Maharashtra," *Man and Environment,* 1, pp. 64-67.

———, 1977b, "On the occurrence of *Hexaprotodon palaeindicus* from the Ghod Valley, Western India," *Jour. Univ. Poona Sci. & Tech.* 50, pp. 209-19.

Badam, G.L. and S.N. Rajaguru, 1991, "Problems of Neogene Litho-and Biostratigraphy of the Siwaliks in Northwest India," *Birbal Sahni Birth Centenary Symposium on Siwalik Basin,* 7-9. Dehradun, Wadia Institute of Himalayan Geology.

Badam, G.L. and Salahuddin, 1982, "Some new fossil sites on the Central Narmadā valley, Madhya Pradesh," *Curr. Sci.* 51 (18), pp. 898-99.

Badam, G.L. and Vijay Sathe, 1992, "Animal depictions in Rock Art and Palaeoecology — A case study at Bhimbetka, Madhya Pradesh, India," in Pager, S-A, B.K. Swartz Jr. & A.R. Willocx (eds.) *Proceedings of the Rock Art — The Way Ahead, International Conference on Rock Art Conservation, Recording and Study,* Natal, South Africa, pp. 196-208.

Badam, G.L., R.K. Ganjoo and Salahuddin, 1986, "Preliminary Taphonomical Studies of Some Pleistocene Fauna from the Central Narmadā Valley, Madhya Pradesh, India," *Palaeogeography, Palaeoclimatology, Palaeoecology* 53, pp. 335-48.

Badam, G.L., R.K. Ganjoo, Salahuddin, and S.N. Rajaguru, 1986, "Evaluation of Fossil hominid — the maker of late Acheulian tools at Hathnorā, Madhya Pradesh, India," *Curr. Sci.* 55 (3), pp. 143-45.

Badam, G.L., V.D. Mishra, J.N. Pal and J.N. Pandey, 1989, "A Preliminary Study of Pleistocene Fossils from the Middle Son Valley, Madhya Pradesh," *Man and Environment*, 13, pp. 41-47.

Behrensmeyer, A.K., 1975, "The Taphonomy and Palaeoecology of Plio-Pleistocene Vertebrate Assemblages, East of Lake Rudolf, Kenya," *Bull. Mus. Comp. Zool*, 146 (10), pp. 473-78.

———, 1990, *Experimental Taphonomy Workshop,* Smithsonian Institution, Washington D.C. International Council for Archaeology.

Behrensmeyer, A.K. and S.M. Kidwell, 1985, "Taphonomy's Contribution to Paleobiology," *Paleobiology,* 11, pp. 105-45.

Biswas, S. and D.C. Dassarma, 1981, "A new species of *Stegodon* from the alluvial deposits of Narmadā Valley in Peninsular India," *Proc. Field Conference Neogene Quaternary Boundary,* 1979, pp. 15-20.

———, 1984, Fossil Hippopotamus from the Quaternary alluvial deposits of the Narmada Valley in M.P. *Rec. Zool. Surv. Ind.,* 114 (2), pp. 19-24.

Blanford, W.T., 1869, "On the Geology of the Taptee and the Nerbudda valley and some adjoining districts," *Memoirs of the Geological Survey of India* 6(3), pp. 163-282.

Clutton-Brock, J., 1965, "Excavations at Langhnaj, 1944-63," *The Fauna,* Deccan College Ser. 27, pp. 1-41.

Colbert, E.H., 1935, "Siwalik mammals in the American Museum of Natural History," *Transactions of the American Philosophical Society* (n.s.) 26, 401 pp.

———, 1942, "The geological succession of the Proboscidea in India," in: Osborn, H.F., *Proboscidea, Amer. Mus. Nat. History,* vol. 2, pp. 1421-1521.

Dassarma, D.C. and S. Biswas, 1978, "Fossil Hippopotamus from Madhya Pradesh," *Geol. Surv. Ind.* News 9 (6), p. 7.

———, 1979, "Some Observations on the Quaternary Stratigraphy and Mammal Assemblages of India," *G.S.I. Misc. Publ.* 45, pp. 279-87.

de Terra, H. and T.T. Paterson, 1939, "Studies on the Ice age in India and associated human cultures," *Publ. Carn. Inst. Washington* 493, pp. 1-354.

Delson, E., 1985, Late Pleistocene Human Fossils and Evolutionary Relationship in: *Ancestors: The Hard Evidence* (ed. E. Delson, New York): Alan R. Liss.Inc, pp. 296-300.

Deraniyagala, P.E.P., 1956, Some extinct Hippopotami and Elephants from India and Ceylon, *Jour. Pal. Soc. India,* 1, pp. 176-78.

Efremov, J.A., 1940, "Taphonomy: A New Branch of Palaeontology," *Pan-American Geologist* 74, pp. 81-93.

Epstein, H. and I.L. Mason, 1984, Cattle, in *Evolution of Domesticated Animals* (I.L. Mason, ed.), London and New York: Longmans, pp. 6-27.

Falconer, H., 1859, *Descriptive Catalogue of the Fossil Remains of Vertebrate from the Siwalik Hills, the Nerbudda, Perim Island, etc. in the Museum of the Asiatic Society of Bengal*, Calcutta.

Falconer, H. and P.T. Cautley, 1836, Note on the Fossil Hippopotamus of the Siwalik Hills. *Asi. Res.*, 19, pp. 39-53.

Gaur, R., 1981, Palaeontology and Palaeoecology of Plio-Pleistocene Upper Sivalik Sediments in the north-east of Chandigarh, PhD Diss. Panjab University Chandigarh.

Gifford, D.P., 1981, "Taphonomy and Palaeoecology: A Critical Review of Archaeology's Sister Disciplines," in *Advance in Archaeological Method and Theory* 4, New York: Academic Press, (M.B. Schiffer, ed.), pp. 365-438.

Gilbert, A.S., 1979, *Urban Taphonomy of Mammalian Remains from the Bronze Age of Godin Tepe, Western Iran*, PhD Dissertation, Ann Arbor: Columbia University.

Grigson, C., 1985, "*Bos indicus* and *Bos namadicus* and the Problem of Autochthonous domestication in India," in Misra V.N. and P. Bellwood, (eds.), *Recent Advances in Indo-Pacific Prehistory*, New Delhi: Oxford and IBH Publishing House, pp. 425-28.

Groves, C.P., 1982, "Geographical variation in the Barasingha or Swamp Deer," *Journal of Bombay Natural History Society of India*, 79 (3), pp. 620-29.

Hooijer, D.A., 1950. The Fossil Hippopotamidae of Asia with notes on the recent species. *Zool. Verhand. Mus. Leiden*, 8, pp. 1-124.

———, 1955, "Fossil Proboscidea from the Malay Archipelago and the Punjab," *Zool. Verhand. Mus. Leiden*, 28, pp. 1-146.

———, 1963, "Preliminary Identification of some Fossil Mammals from Narmadā Valley, India," *Rivista Di Science Prehistoric* 18 (1-4), p. 20.

Hopwood, A.T., 1935, Fossil Elephants and Man, *Proc. Geol. Assoc. London*, 46, pp. 46-60.

Joglekar, P.P. and P.K. Thomas, 1990, "Ancestry of *Bos* Species: Myth and Reality: I. Genetic Evidence," *Man and Environment*, vol. 15 No.1, pp. 51-55.

Joshi, R.V., G.L. Badam and R.P. Pandey, 1978, "Fresh data on the Quaternary Animal Fossils and Stone Age Cultures from the Central Narmadā Valley, India," *Asian Perspectives*, 21 (2), pp. 164-81.

Kennedy, K.A.R. and J. Chiment, 1991, The fossil hominid from the Narmadā Valley, India": *Homo erectus* or *Homo sapiens*? in: indo-Pacific Prehistory 1990, edited by P. Bellwood, 1, pp. 42-50, Canberra, Indo-Pacific Prehistory Association.

Kennedy, K.A.R., J. Chiment Disotell, T. and Meyers, D., 1991, "Principal Components Analysis of Prehistoric South Asian Crania," *American Journal of Physical Anthropology* 64, pp. 105-18.

Khatri, A.P., 1961, "Stone Age and Pleistocene Chronology of the Narmadā Valley," *Anthropos* 56, pp. 519-30.

———, 1963, "Mahadevian: An aldowan pebble Culture of India," *Asi. Persp.*, 6, pp. 186-97.

———, 1966, "The Pleistocene Mammalian Fossils of the Narmadā River Valley and their Horizons," *Asian Perspectives*, 9, pp. 113-33.

Kshirsagar, Anupama and G.L. Badam, 1990, "Biochronology and Fluorine analysis of some Pleistocene fossils from Central and Western India," *Bull. Deccan College PGRI*, 49: pp. 199-211.

Kumar, Arun, 1985, *Quaternary Studies of the Upper Godavari Valley* (*A Study in Environmental Archaeology*), Poona Univ.: Poona, Unpublished PhD Thesis.

Kumar, Sanjay, 1999, *Taphonomy and Environment of Pleistocene Fossils from the Central Narmadā Valley, Narsinghpur District, M.P.*, M.A. Dissertation, Deccan College, Pune.

Lydekker, R., 1880, "Siwalik and Narbada Proboscidea," *Pal. Ind.* 10 (5), pp. 182-292.

———, 1882, Notes on some Siwalik and Jamuna Mammals, *Rec. Geol. Surv. Ind.*, 15, 28-33.

———, 1884, "Siwalik and Narbada Bunodont Suina," *Pal. Ind.* 3 (2), pp. 35-104.

———, 1886, *Catalogue of Fossil Mammalia in the British Museum*, pts. 3 & 4, London.

Martin, C., 1977, "Status and Ecology of the Barasingha (*Cervus duvauceli branderi)* in Kanha National Park, India," *Journal of Bombay Natural History Society* 74 (1), pp. 61-132.

Matthew, W.D., 1929, "Critical Observations upon Siwalik Mammals," *Bull. Ameri. Mus. Nat. Hist.* 56, pp. 437-560.

Medlicott, H.B., 1873, Note on the celt found by Mr. Hacket in the ossiferous deposits of the Narmadā Valley (Pliocene of Palconer); on the age of the deposits and the associated shells, *Rec. Geol. Surv. India.*, 6(3), pp. 49-54.

Medlicott, J.G., 1860, "On the geological structure of the central portion of the Nerbudda District," *Memoirs of the Geological Survey of India* 2 (2), 118 pp.

Minton, S.A., 1966, A Contribution to the Herpetology of West Pakistan. *Bull. Amer. Mus. Nat. Hist.*, 134, pp. 59-70.

Mishra, S. and S.N. Rajaguru, 1993, Quaternary deposits of Bheḍaghāṭ, Jabalpur, *Man & Environment*, 28, pp. 10-12.

Nanda, A.C., 1978, *Hexaprotodon sivalensis* Falconer and Cautley from the Upper Siwalik Sub-group, Ambala, Haryana. In: *Rec. Res. Geol.* (V.J. Gupta, ed.), Hindustan Publishing Corporation, Delhi, 5, pp. 161-74.

Oldham, R.D., 1893, *A Manual of the Geology of India*, 1st edn., Calcutta: Superintendent of Government Printing.

Oldham, T., 1871, "Sketch of the geology of the Central Provinces," *Records of the Geological Survey of India* 4(3), pp. 69-81.

Osborn, N.F., 1942, "Proboscidea. 2, Stegodontoiden Elephantoidea," *Amer. Mus. Nat. Hist.*, pp. 805-1676.

Owen, R., 1845, *Odontography*, London.

Patnaik, Rajeev, 1995, "Narmadā Valley Microvertebrates: Systematics, Taphonomy and Palaeoecology," *Man and Environment*, 22 (2), pp. 75-90.

Patnaik, Rajeev, G.L. Badam and Vijay Sathe, 1995, "Discovery of Microvertebrates from the Pleistocene deposits of the Central Narmadā Valley, India," *Curr. Sci.* 68(8), pp. 828-30.

Pilgrim, G.E., 1905, "On the occurrence of *Elephas antiquus* (*namadicus*) in the Godavari alluvium with remarks on the species, its distribution and the age of the associated Indian deposits," *Rec. Geol. Surv. India* 32 (3), pp. 199-218.

———, 1939, "The fossil Bovidae of India," *Mem. Geol. Surv. India. Palaeont, Indica, N.S.*, 26, 356 pp.

Princep, J., 1832, "Note on the Jabalpur Fossil bones," *Journ. Asi. Soc. Bengal*, 1, pp. 456-58.

———, 1833, "Note on the fossil bones discovered near Jabalpur," *Journal of the Asiatic Society of Bengal* 2, pp. 583-88.

Sahni, M.R. and E. Khan, 1961, On an abnormal dentition of *Hexaprotodon sivalesis,* Fal & Caut. *Res. Bull. (N.S.) Panj. Univ., Chandigarh,* 12, pp. 257-58.

Salahuddin, 1989, *Late Quaternary Ecology, Fauna and Human Cultures of the Central Narmadā Valley, Madhya Pradesh,* unpublished PhD Thesis, Poona University.

Salahuddin and G.L. Badam, 1983, "A note on the discovery of Upper Palaeolithic culture from the Central Narmadā Valley, Madhya Pradesh," *Curr. Sci.* 52 (20), pp. 976-78.

Salahuddin, G.L. Badam, R.K. Ganjoo and S.N. Rajaguru, 1986-87, "On the Archaeological Association of the fossil Hominid from Hathnorā, Madhya Pradesh, India," *Asian Perspectives*, 27 (2), pp. 193-203.

Sankalia, H.D., 1974, *Prehistory and Protohistory of India and Pakistan*, Pune: Deccan College.

Saxena, M.N., 1968, "A Note on the Validity of the Upper Siwalik Vertebrates Fossils for Stratigraphic Correlation," *The Eastern Anthropologist* 21 (3), pp. 319-21.

Sen, D. and A.K. Ghosh, 1963, "Lithic culture complex in the Pleistocene sequence of Narmadā Valley, Central India," *Rivista Die Scienze Preistoriche*, 18 (1-4), pp. 3-23.

Sonakia, A., 1984, "The Skull Cap of Early Man and Associated Mammalian Fauna from Narmadā Valley Alluvium, Hoshangabad Area, M.P. (India)," *Rec. Geol. Surv. India:* 113, pp. 159-72.

———, 1985, "Skull Cap of an Early Man from the Narmadā Valley Alluvium (Pleistocene) of Central India," *American Anthropologist*, 87 (3), pp. 612-16.

Spilsbury, G.G., 1833, "Account of the fossil bones discovered in the bed of the Omar Nadi near Narsinghpur on Gadarwara in the valley of the Nerbudda," *Journal of the Asiatic Society of Bengal* 2, pp. 388-95.

———, 1837, "Notice of new sites of fossil deposits in the Nerbudda valley," *Journal of the Asiatic Society of Bengal* 6, pp. 487-89.

———, 1844, "Notes on Nerbudda fossils," *Journal of the Asiatic Society of Bengal*, 13, pp. 765-66.

Sukumar, R., 1989, *The Asian Elephant : Ecology and Management*, Cambridge: Combridge University Press.

Supekar, S.G., 1968, *Pleistocene Stratigraphy and Prehistoric Archaeology of the Central Narmadā Basin*, Unpublished PhD thesis, Poona University.

Tewari, B.S. and G.L. Badam, 1969, A new Species of fossil turtle from the Upper Siwaliks of Pinjore, India. *Palaeont.* 12(4), pp. 535-58.

Theobald, W., 1860, "On the Tertiary and alluvial deposits of the central portion of the Nerbudda valley," *Memoirs of the Geological Survey of India* 2, pp. 279-98.

Tiwari, M.P. and H.Y. Bhai, 1997, "Quaternary Stratigraphy of the Narmadā Valley," *Geol. Surv. Ind. Spl. Pub.* 46, pp. 33-63.

Zeuner, E.F., 1950, *Dating the Past*, 2nd edn., London: Meuthen.

Index

Note: Index of species, fossil localities, geological formations, lithounits, geographic names (places, rivers, mountains, etc.) has been included. Only the species names have been indexed in case of appendices.